Simulation and Power Analysis Using R

Simulation-based methods are increasingly central to modern applied research, particularly as study designs and statistical models grow more complex. Traditional analytic tools for power analysis and study planning often rely on simplifying assumptions that are difficult to justify in real-world settings. Simulation provides a flexible alternative, allowing researchers to explore design choices, modeling assumptions, and inferential tradeoffs under realistic data-generating processes. ***Simulation and Power Analysis Using R*** presents simulation not only as a tool for estimating power, but as a broader framework for study design and causal reasoning. Using applied examples and clear conceptual motivation, the book demonstrates how simulation can inform decisions before data are collected.

Features

- Practical and applied introduction to simulation-based power analysis and study design
- Coverage of generalized linear models and generalized linear mixed models, including clustered and hierarchical data
- Extensions to experimental and quasi-experimental designs such as randomized controlled trials, cluster randomized trials, regression discontinuity, difference-in-differences, and interrupted time series
- Strategies for varying simulation parameters to explore robustness and design tradeoffs
- Methods for simulating realistic data conditions, including missing data and model misspecification
- Fully reproducible R code throughout, built around a modular simulation framework using the simglm package

This book is intended for applied researchers, data scientists, and graduate students in fields such as statistics, education, psychology, public health, and the social sciences. It is particularly well suited for researchers who need to justify study design or power calculations in grant proposals, or who work with complex designs for which standard formulas are inadequate. The book may also be used as a text for graduate courses or advanced seminars on simulation, power analysis, or applied causal inference.

Brandon LeBeau is a data scientist, statistical software developer, consultant, and educator with expertise in applied statistical methodology, causal inference, program evaluation, research software development, and simulation-based statistical methods. He has developed multiple research software packages available on CRAN and GitHub, including the simglm R package (LeBeau, 2025), which serves as the primary simulation framework used throughout this book.

Chapman & Hall/CRC
The R Series

Series Editors

Recently Published Titles

Spatial Data Science: With Applications in R
Edzer Pebesma and Roger Bivand

Modern Data Visualization with R
Robert Kabacoff

Learn R: As a Language, Second Edition
Pedro J. Aphalo

Spatial Analysis in Geology Using R
Pedro M. Nogueira

Analyzing Baseball Data with R, Third Edition
Jim Albert, Benjamin S. Baumer and Max Marchi

Geocomputation with R, Second Edition
Robin Lovelace, Jakub Nowosad, Jannes Muenchow

Microeconometrics with R
Yves Croissant

Statistical Inference via Data Science
A ModernDive into R and the Tidyverse, Second Edition
Chester Ismay, Albert Y. Kim, and Arturo Valdivia

Copula Additive Distributional Regression Using R
Giampiero Marra and Rosalba Radice

Introduction to Political Analysis in R
H. Whitt Kilburn

Displaying Time Series, Spatial, and Space-Time Data with R
Oscar Perpiñán Lamigueiro

Interactively exploring high-dimensional data and models in R
Dianne Cook and Ursula Laa

Simulation and Power Analysis Using R
Brandon LeBeau

For more information about this series, please visit: https://www.crcpress.com/Chapman--HallCRC-The-R-Series/book-series/CRCTHERSER

Simulation and Power Analysis Using R

Brandon LeBeau

CRC Press
Taylor & Francis Group
Boca Raton London New York

CRC Press is an imprint of the
Taylor & Francis Group, an **informa** business
A CHAPMAN & HALL BOOK

First edition published 2027
by CRC Press
2385 NW Executive Center Drive, Suite 320, Boca Raton FL 33431

and by CRC Press
4 Park Square, Milton Park, Abingdon, Oxon, OX14 4RN

CRC Press is an imprint of Taylor & Francis Group, LLC

Library of Congress Cataloging-in-Publication Data
Names: LeBeau, Brandon C. author
Title: Simulation and power analysis using R / Brandon LeBeau.
Description: First edition. | Boca Raton, FL ; Abingdon, Oxon : CRC Press, 2027. | Series: Chapman & Hall/CRC the R series | Includes bibliographical references and index. | Summary: "Data simulation is a popular method used to explore statistical methods to understand their properties when data do not meet the statistical assumptions of the method. This is especially important when the models become more complex as mathematical derivations increase significantly. Data simulation is a way to make data assumptions that could better mimic data that would be collected in the real-world and understand how well the statistical analysis method performs under these data conditions. This book provides an overview of methods for data simulation of generalized linear mixed models with R using the simglm package developed by the author"-- Provided by publisher.
Identifiers: LCCN 2026013190 (print) | LCCN 2026013191 (ebook) | ISBN 9781032012506 hardback | ISBN 9781032012513 paperback | ISBN 9781003177876 ebook
Subjects: LCSH: Statistical hypothesis testing--Data processing | Statistical power analysis--Data processing | R (Computer program language) | Computer simulation
Classification: LCC QA277 .L4248 2027 (print) | LCC QA277 (ebook)
LC record available at https://lccn.loc.gov/2026013190
LC ebook record available at https://lccn.loc.gov/2026013191

ISBN: 978-1-032-01250-6 (hbk)
ISBN: 978-1-032-01251-3 (pbk)
ISBN: 978-1-003-17787-6 (ebk)

DOI: 10.1201/9781003177876

Typeset in Latin Modern font
by KnowledgeWorks Global Ltd.

Publisher's note: This book has been prepared from camera-ready copy provided by the authors.

To my family.

Contents

Preface

Why read this book

This book will explore reasons for why simulation and power analysis by simulation can be useful for a variety of problems. The focus of the book will be less on the theory of simulation or power analysis, but rather, practical ways to simulate and conduct power by simulation using the **simglm** R package. As such, this book may be useful for graduate students in quantitative disciplines, researchers planning quantitative studies, or others looking to simulate data for class or research. To my knowledge, data simulation is often a skill that is not directly taught in many graduate programs and therefore can be an almost mystical process. This book hopes to provide practical examples to help new users who wish to simulate data or conduct power analysis by simulation start on that journey.

Why use the **simglm** R package for the book? For one, I wrote the package, and secondly, it attempts to provide a framework that one can build off of for many different types of research design or statistical analysis. The framework is flexible where users can start simple and build more complex situations by adding more pieces to the simulation or power analysis pipeline.

Structure of the book

Chapter 1 will provide an introduction to why users may want to use simulation in their work. Chapters 2 through 6 will cover the basics of simulating data using the **simglm** R package. These chapters start with simulating data based on a linear regression model, then proceed to other types of simulation generating models including linear mixed models and generalized linear mixed models.

Chapters 7 through 10 will introduce users to power analysis and the benefits of estimating power by simulation. Within these chapters, the **simglm** syntax for power analysis by simulation will be explored. As part of this discussion, some best practices for power estimation using simulation will be presented.

Lastly, Chapters 11 through 17 will conduct case studies of power estimation by simulation for common research designs. Within these chapters, the power estimation will follow directly from a published empirical study of that design. The published empirical studies will be used as a baseline to show how to generate realistic data conditions to estimate power by simulation for those types of research designs. Commonly used experimental and quasi-experimental designs will be used as the primary research designs. The research designs that will be explored include:

- Randomized Controlled Trials
- Cluster Randomized Controlled Trials
- Interrupted Time Series
- Regression Discontinuity Design
- Difference in Differences
- Propensity Scores
- Non-Experimental Designs

Software information and conventions

I used the **knitr** package (Xie, 2015) and the **bookdown** package (Xie, 2022) to compile this book. My R session information is shown below:

```
xfun::session_info()
```

```
## R version 4.5.0 (2025-04-11)
## Platform: aarch64-apple-darwin20
## Running under: macOS 26.2, RStudio 2025.5.0.496
##
## Locale: en_US.UTF-8 / en_US.UTF-8 / en_US.UTF-8 / C / en_US.UTF-
8 / en_US.UTF-8
##
## Package version:
##   base64enc_0.1.6   bookdown_0.46
##   bslib_0.10.0      cachem_1.1.0
##   cli_3.6.5         compiler_4.5.0
##   digest_0.6.39     evaluate_1.0.5
##   fastmap_1.2.0     fontawesome_0.5.3
##   fs_1.6.6          graphics_4.5.0
##   grDevices_4.5.0   highr_0.11
##   htmltools_0.5.9   jquerylib_0.1.4
##   jsonlite_2.0.0    knitr_1.51
##   lifecycle_1.0.5   memoise_2.0.1
```

```
##    methods_4.5.0      mime_0.13
##    otel_0.2.0         R6_2.6.1
##    rappdirs_0.3.4     rlang_1.1.7
##    rmarkdown_2.30     rstudioapi_0.18.0
##    sass_0.4.10        stats_4.5.0
##    tinytex_0.58       tools_4.5.0
##    utils_4.5.0        xfun_0.56
##    yaml_2.3.12
```

The book will use the following conventions regarding R code in the text. Package names are shown in bold text (e.g., **rmarkdown**), and inline code and filenames are formatted in a typewriter font (e.g., `knitr::knit('foo.Rmd')`). Function names are followed by parentheses (e.g., `bookdown::render_book()`). Larger code chunks will have a light gray background.

Acknowledgments

I first want to thank my family, Rachel, Amelie, and Emmeline, for providing love and support throughout the writing process. I couldn't have a more committed, loving, silly family; I love you all. I also thank my parents and my grandma, Carol LeBeau, who kept pushing me to achieve my potential; I'm indebted to that support, which set me up to succeed.

I also want to thank Andrew Zieffler for getting me really started down the R path many years ago and my graduate school advisor, Michael Harwell, who supported my dissertation plan. That dissertation[1] was where I had the initial structure of the **simglm** package and first working prototype. Lastly, I thank Ariel Aloe, a great colleague and friend, who has greatly influenced my thinking on many topics.

[1] https://conservancy.umn.edu/items/8ce8b075-dfde-4f1b-9371-1e6932db4229

About the Author

Brandon LeBeau is a data scientist, statistical software developer, consultant, and educator with expertise in applied statistical methodology, causal inference, and reproducible modeling. His research interests include quantitative program evaluation, research software development, and simulation-based statistical methods. He has developed multiple research software packages available on CRAN and GitHub, including the simglm R package (LeBeau, 2025), which serves as the primary simulation framework used throughout this book.

1

Introduction

1.1 Chapter Overview

This chapter introduces simulation as a unifying framework for study design, methodological decision-making, and causal reasoning in applied research. Rather than treating statistical methods as fixed recipes, the chapter motivates simulation as a way to explore tradeoffs among design choices, assumptions, and analytical strategies before data are collected. The role of simulation is framed broadly, extending beyond power analysis to include sensitivity analysis, model comparison, and the evaluation of design robustness under realistic data-generating processes. The chapter outlines the goals of the book, the intended audience of applied researchers, and the types of research questions for which simulation-based approaches are particularly valuable. It also provides an overview of the modeling frameworks and study designs covered throughout the book, including randomized experiments, quasi-experimental designs, multilevel models, and generalized linear models. Emphasis is placed on transparency, reproducibility, and interpretability, setting the stage for later chapters that integrate simulation with causal inference and applied data analysis.

This book introduces simulation and power by simulation using R. In particular, the book focuses on a single R package, **simglm**, to present a unified framework for simulation and power-by-simulation analyses. This package is flexible and allows for the simulation of linear (mixed) models, including generalized linear (mixed) models.

Simulation and power by simulation have become much more common in recent years for many reasons. First, software has become much more usable by a broader audience to engage with these methods. Second, simulation-based approaches have been shown to be useful for evaluating common assumption violations within power estimation and for assessing the impact those violations have on estimation and power (Morris et al., 2019). Finally, computational power, especially cloud computing, has made it possible to obtain more precise estimates from simulation and power-by-simulation analyses by performing a larger number of iterations more efficiently.

DOI: 10.1201/9781003177876-1

1.2 What Is Simulation?

Simulation is the process of generating synthetic data. Synthetic data are often generated to mimic data collected in the real-world. There are many ways to generate synthetic data, and it is beyond the scope of this book to cover all such approaches. This book focuses on one specific, albeit broad, way to generate synthetic data. The book utilizes the generalized linear (mixed) models to generate synthetic data.

This model is flexible to accommodate many different situations and outcomes. For example, the generalized linear (mixed) model can simulate synthetic data that would mimic the following situations or research designs:

- A two (or more) group design to evaluate the mean differences between groups,
- Longitudinal or time series data,
- Clustered data, for example, patients nested within hospitals or students nested within schools,
- Non-linear trajectories of attributes.

Furthermore, the generalized linear (mixed) model can simulate the following types of outcomes (and more):

- continuous outcomes,
- ordinal or discrete outcomes with many levels,
- binary outcomes (e.g., mastering a skill versus not mastering a skill),
- count outcomes (i.e., number of times an event occurred).

When different research designs are combined with different outcome types, the variety of synthetic data that can be generated expands greatly. More details on the specific steps to simulate the synthetic data from the generalized linear (mixed) model are introduced beginning in Chapter 2.

1.2.1 Usefulness of Simulation

Simulation can be used by a variety of stakeholders. Simulation could be used to show a particular method within a classroom context. For example, instead of using real-world data, data can be simulated to fit the specific context that is being discussed in class. This can give the instructor control about exactly what and how the data are behaving for instructional benefits. Beyond instruction, simulation can be used to evaluate how sensitive analytic methods are to the characteristics of the data being collected, such as heterogeneity or other departures from idealized assumptions. For example, if the data are heterogeneous, how do different methods perform? This can be important for avoiding exploratory "fishing" within observed real-world data when addressing a particular research question.

Finally, as discussed further in the next section, simulation can be used as a way to estimate power for study planning. Most commonly, this is useful for grant preparation to show that the number of individuals to be recruited is sufficient to detect a specific magnitude of the effect. For example, if it is hypothesized that the population mean difference between two groups is 0.25 standard deviations, this approach allows researchers to evaluate how likely the proposed analytic method is to accurately estimate this effect given the planned sample. Power by simulation can be a good alternative to traditional methods that rely on strong assumptions about the data to be collected. Often, all of these assumptions will not be met with real data and power by simulation can provide a way to determine how sensitive the results will be if the assumptions are not met. This will be discussed in more detail starting in Chapter 7.

1.3 Why Perform Power by Simulation?

Estimating power by simulation can enhance power arguments by aligning the data generation process with data conditions that are likely to be collected in the field. For example, building in treatment heterogeneity, missing data, and other conditions can be flexibly explored to understand their impact on power. These model-based assumptions do not need to be directly assumed when estimating power through simulation compared to traditional closed-form equations.

Power by simulation can also take advantage of the Monte Carlo simulation process to directly estimate the sampling distribution of the statistic of interest. This can be helpful when different statistics or models that do not have nice closed-form equations are desired. Having an estimate of the sampling distribution of the statistic can also allow for a shift in mindset to reframe power as the likelihood of seeing a practically useful estimated effect size. This would be different from the null hypothesis significance testing framework where power represents the likelihood of rejecting a null hypothesis that the population effect is equal to zero. This hypothesis is typically not as interesting when evaluating the power due to researchers having strong justification that the population effect size is likely not zero. This creates a mismatch between the primary statistical test and researchers' prior knowledge about the expected impact of the treatment.

1.4 Simulation and Causal Inference

A major focus of this book is the use of simulation to support causal inference in experimental, quasi-experimental, and non-experimental research designs. Causal inference methods aim to estimate the effect of an intervention, treatment, or exposure. In experimental methods, random assignment is possible, whereas in quasi-experimental or non-experimental settings, random assignment is not performed or is not feasible. In these settings, identification relies heavily on design assumptions rather than on statistical modeling alone.

Simulation provides a natural framework for exploring causal inference methods because it allows researchers to explicitly encode the data-generating process implied by a causal research design. By simulating data under known conditions, researchers can evaluate how well different analytic strategies recover causal effects when key assumptions are met, approximately met, or violated. This makes simulation particularly well suited for studying causal designs that depend on assumptions such as parallel trends, conditional ignorability, continuity at a cutoff, or stable treatment effects over time.

In the second half of this book, simulation is used to study a range of commonly used causal inference approaches, including randomized controlled trials, propensity score methods, interrupted time series designs, difference-in-differences, regression discontinuity designs, and related extensions. Rather than treating these methods as purely analytic tools, the emphasis is placed on understanding their performance as a function of the underlying data-generating process. This perspective highlights the close connection between study design, identifying assumptions, and statistical power.

By framing causal inference methods through simulation, the book emphasizes that power, bias, and precision are properties of a design operating under specific conditions, not fixed characteristics of a statistical test. This approach allows researchers to more realistically evaluate whether a proposed study design is capable of answering a causal question of interest. This book is not intended to teach the specifics of causal inference methods; rather, it uses these methods as vehicles for estimating power under realistic causal designs. Readers interested in a more comprehensive treatment of causal inference are directed to Cunningham (2021), Hernán and Robins (2020), Huntington-Klein (2021), Imbens and Rubin (2015), Lawson (2014), Morgan and Winship (2014), Mosteller and Boruch (2002), Pearl (2009), and Rubin (1974) for more information on these study designs.

This book aims to bridge experimental, quasi-experimental, and observational methodologies by unifying them within a simulation-based framework. The focus is not on simulation as a purely computational exercise, but on its use as a design-oriented tool that clarifies assumptions, illuminates trade-offs, and supports principled study planning across a range of causal settings. Throughout the book, simulation is used to connect statistical models to the underlying data-generating processes they are intended to represent.

1.5 Why Read This Book?

This book is intended for researchers who are interested in simulation or power by simulation. This book assumes readers are familiar with generalized linear (mixed) models. Some basics of these models will be presented, but going into thorough details on these models is beyond the scope of this book. Those interested in a refresher on these topics are directed to Fitzmaurice et al. (2012), Fox (2015), or Gelman et al. (2020).

The book does not assume any prior knowledge of simulation. The steps to simulate synthetic data from the generalized linear (mixed) models will be built up from scratch. Finally, it is expected that readers are familiar with the R statistical language as this is the language used for the simulation and power by simulation. If readers would like some background on R, the R for Data Science (Wickham and Grolemund, 2016) book is recommended as a good introduction to R for data science with the tidyverse (Wickham et al., 2019). The **simglm** package used throughout the book bases the coding style off of the tidyverse to build the pieces of the simulation. The package can be used without direct knowledge of the tidyverse.

2

Simulating General Linear Models

2.1 Chapter Overview

This chapter introduces simulation of data from general linear models as a foundation for simulation-based study design and power analysis. The chapter focuses on specifying data-generating processes that reflect common regression settings encountered in applied research, including continuous outcomes, multiple predictors, and correlated covariates. Key components of linear model simulation are discussed, such as defining regression coefficients, controlling error variance, standardizing predictors, and interpreting effect sizes. The chapter demonstrates how simulation parameters map directly onto population quantities and how those parameters influence statistical power and estimation accuracy. Practical examples illustrate how simulated data can be used to explore the consequences of sample size, model misspecification, and effect magnitude. Throughout, the emphasis is on helping applied researchers understand how linear models behave under known conditions and how simulation can be used to answer design questions that are difficult to address analytically.

General linear models (LM) take data attributes and relate them to some outcome of interest. The general linear model is sometimes referred to as the special case, linear regression. This model takes the following general matrix form:

$$\mathbf{Y} = \mathbf{X}\beta + \epsilon \tag{2.1}$$

where $\mathbf{Y}$ is a vector representing the outcome attribute, $\mathbf{X}$ is a matrix of data attributes representing fixed effects, β is a vector of regression coefficients, and ϵ represents random error. It is beyond the scope of this book to give a fully in-depth introduction and description of these models. A brief description of these terms follows to understand better how to simulate data from a model like this and how these correspond to the functions used in the **single** package. Those interested in a more in-depth introduction to the LM model can see (Faraway, 2004; Fox, 2015; Gelman et al., 2020; Lewis-Beck and Lewis-Beck, 2015).

DOI: 10.1201/9781003177876-2

2.1.1 Understanding the Basics of Linear Regression Terms

The fixed effects in a linear regression are assumed to be measured perfectly (i.e., without measurement error) and are also assumed to be fixed. This is where the fixed effect moniker stems from and refers to the values in the data, which are the only ones we are interested in making inferences about. More specifically, consider a situation where the types of trees were collected from a single region as part of broader data on forestry. In this example, all the different types of trees native to a particular region could be collected. In addition, these would represent all possible native tree species for that region, indicating that these would be the only tree types we would want to make conclusions about. This idea of a fixed effect could be applied to any other data collected.

The fixed effect name comes from experimental settings where the elements of an experiment could be replicated perfectly (Fox, 2015). However, this is not often the case in social science or applied research. For example, consider the case where a sample of data is collected on depressive symptoms for an individual. In this situation, sample data likely does not contain all possible values of depressive symptoms that could occur. Therefore, some data from the theoretical range may need to be included.

2.2 Simulating the General Linear Model

The steps to simulating from the general linear model in the simplest case can be broken down into six steps. These steps are written in words below with some indication about what function when using the **simglm** package; each will be discussed in more detail:

1. Specify the terms for the model; for example, how many fixed terms are there?
2. Define and specify the simulation terms for the fixed attributes.
3. Define and specify the simulation terms for the random error.
4. Define the relationship between the fixed attributes ($\mathbf{X}$) and the outcome attribute ($\mathbf{Y}$), that is the β weights.
5. Based on steps 2 through 4, simulate
    - the fixed terms (e.g., using `simulate_fixed()`)
    - the random error (e.g., using `simulate_error()`)
6. Generate the outcome based on steps 2 through 5 (e.g., using `generate_response()`).

These steps will be used throughout the book; therefore, let us think about each step. Throughout these descriptions, **simglm** code will be used to build the simulation arguments and simulate the data. After the description of each, a complete example model will be simulated.

2.2.1 Specify Model Terms

This step determines the model and scope of data to be simulated. It is the least technical stage as it does not require any code (although the model will be written out in R equation code), but it is a critical step. As such, an entire chapter (Chapter 6) is dedicated to thinking through how to simulate data that are realistic of real-world data conditions. Simulating realistic data conditions is important for at least three reasons: realistic data would 1) better inform the reality, 2) better estimate power for the desired example, and 3) may generalize to more data situations. This section will focus on the specification of the model formula for generating the population model.

Specifying the model is done through articulating what will impact the outcome of interest. For example, if the outcome of interest is academic achievement or the amount of algae in a water ecosystem, the model's specification would define what impacts or explains variation in that outcome. In the example of academic achievement, other data characteristics could include prior academic achievement, socioeconomic status (SES), or other data characteristics. This step does not need to include the exact relationship between the explanatory data characteristics (e.g., SES) and the outcome of interest; instead, it just includes which attributes help to explain variation in the outcome. The average relationship will be specified later in the simulation process.

Within the **simglm** package, a list of simulation arguments is created prior to doing the actual simulation. For this step, the population model formula needs to be specified, which is done as follows:

```
library(simglm)

sim_arguments <- list(
  formula = achieve ~ 1 + prior_achieve + SES + engagement +
    motivation + state
)
```

In the code above, the simulation arguments are starting to be built to simulate the data. These are being stored in the object, `sim_arguments`, but any name can be used for these. Within this object, a list is being built. This list needs specific characteristics for the **simglm** package to process them properly. Specifically, for this stage, the named element of the list, `formula`, is used to define the population model. The `formula` argument of the list takes a two-sided formula that follows typical R formula specification where the outcome is on the left-hand side of the ~ and any predictor attributes are on the right-hand side. This model formula could be read in words as achievement (`achieve`) is predicted, modeled, or explained by an intercept (`1`), prior achievement (`prior_achieve`), SES (`SES`), parent engagement (`engagement`), motivation (`motivation`), and state of residence (`state`). To match with the linear regression equation shown above, the `achieve` attribute in the formula would be the **Y** above. The other attributes on the right-hand side of the ~ would represent the **X** or fixed effects.

Of note, no other details need to be specified for this model specification step. The specific details for the values of each element on the right-hand side of the equation, that is, the predictor attributes, will be defined in the next step. In a linear regression model, these would commonly be specified as the fixed effects or the **X** terms, also referred to as the design matrix.

2.2.2 Define and Specify Fixed Terms

In this step, the fixed effects from the model formula are defined in more detail. Practically, this is the step that generates the specific data values for the terms defined on the right-hand side of the formula from the previous step, the fixed effects. In order to generate the actual data values for the fixed effects, first, the type of attribute each one represents needs to be specified. This roughly coincides with the levels of measurement that are sometimes discussed in an introductory statistics course: nominal (e.g., categories), ordinal (e.g., ordered categories), and continuous (e.g., integer or ratio) data.

The simulation arguments using the **simglm** package are specified within the `fixed` element of the named list. The `fixed` element is also a named list where each element of the list is specified by the name of each fixed attribute defined in the generating model formula. The structure for defining the fixed element of the named list is shown below, building off the example already started.

```
sim_arguments <- list(
  formula = achieve ~ 1 + prior_achieve + SES + engagement +
    motivation + state,
  fixed = list(
    prior_achieve = list(),
    SES = list(),
    engagement = list(),
    motivation = list(),
    state = list()
  )
)
```

Within each attribute defined in the fixed simulation arguments, a `var_type` argument is used to identify which data attribute should be simulated. Currently, five different values can be specified within the `var_type` argument: `'factor'`, `'ordinal'`, `'continuous'`, `'time'`, and `'knot'`. In the code chunk below, each of the five attributes considered in the generating population model is defined as a specific type. Of note, these are somewhat arbitrarily chosen here and should reflect how the attribute of interest would be measured during the data collection phase of the research. For example, it would be possible to measure a family's socioeconomic status (SES) continuously (e.g., using factor analysis to generate a latent SES score), but the specification below as a factor attribute assumes that these would represent categories. *The SES attribute would likely be ordinal; however, these are defined within the* ***simglm*** *package as integer attributes.* These would represent data that can only take on whole integer values (i.e., no decimal places).

```
sim_arguments <- list(
  formula = achieve ~ 1 + prior_achieve + SES + engagement +
    motivation + state,
  fixed = list(
    prior_achieve = list(var_type = 'continuous'),
    SES = list(var_type = 'factor'),
    engagement = list(var_type = 'ordinal'),
    motivation = list(var_type = 'continuous'),
    state = list(var_type = 'factor')
  )
)
```

Each different type of attribute requires different simulation arguments to help define how the data should be simulated. The three most common types will be explored below in more detail. The final two attribute types currently supported, `time` and `knot` will be explored in other chapters where they may be more commonly used. In particular, these two elements may be most relevant for interrupted time series designs and will be revisited when this design is considered.

2.2.2.1 Factor / Category Simulation

Factor and category attribute simulation is used when `var_type = 'factor'`. The only additional required simulation argument used for this type of attribute simulation is the `levels` argument. Optional arguments include `replace` and `prob`, which will also be discussed. Procedurally, the `sample()` function is used internally when factor or categorical data attributes are simulated. Therefore, any valid arguments to the `sample()` function can be specified for factor or categorical data attributes.

The `levels` argument is used to specify what the factor or categories would represent for that attribute. For example, when considering the SES attribute specified earlier, the categories in the levels argument may be Low, Medium, or High. The following code chunk specifies the level arguments for the SES and state attributes.

```
sim_arguments <- list(
  formula = achieve ~ 1 + prior_achieve + SES + engagement +
    motivation + state,
  fixed = list(
    prior_achieve = list(var_type = 'continuous'),
    SES = list(var_type = 'factor',
               levels = c('Low', 'Medium', 'High')),
    engagement = list(var_type = 'ordinal'),
    motivation = list(var_type = 'continuous'),
    state = list(var_type = 'factor',
                 levels = c('IA', 'CA'))
  )
)
```

Notice above that for the SES attribute, three values will be possible to be simulated/sampled (Low, Medium, and High), and for the state attribute, two values will be possible to be simulated/sampled (IA and CA, representing Iowa and California states within the United States).

It is important to discuss the default behavior of the optional arguments, `replace` and `prob`. By default, `replace = TRUE` means that the attribute levels will be sampled with replacement. More simply, this means that the specified levels can show up in the data more than once. This is likely the desired behavior, but it may not be if there are more levels than the number of rows being generated. To be more explicit, suppose a case where the levels argument had 1,000 different values, and there were only 100 values to be generated. In this case, `replace = FALSE` would be possible without an error, where the sampling would occur without replacement. This would mean that as one of the levels was drawn (i.e., sampled), they would not be put back in to be possible to draw again. This would result in 100 unique values out of the 1,000 possible values. Again, for most applications, this would not represent the desired behavior.

The `prob` argument may be one that users would want to change. The default behavior is for the levels to have an equal likelihood of being sampled or selected, often referred to as a balanced design in statistics. Therefore, the likelihood of each level being drawn would be computed as $1/\#\ of\ levels$, where the one is divided by the number of levels specified. In the case of the SES data attribute, each level of the attribute would likely be sampled as: $1/3$. For the state attribute, each level would have a likelihood of 0.50. Since the `sample()` function is used internally, the data will likely not be balanced for each data generation, but this will occur in the long run or in the aggregate.

TABLE 2.1 Probability associated with each state using the prob argument.

State	Likelihood
IA	0.25
CA	0.75

The `prob` argument would be how this could be altered from the default behavior. For example, if it was hypothesized that more individuals from California would be in the data, then this level could be given a higher likelihood of being sampled. The `prob` argument is a vector of proportions the same length as the number of levels, where each value specified is the likelihood of selecting/sampling that element of the levels vector. In addition, the sum of the proportions specified needs to add to one. To be more explicit, suppose that we felt California was likely to make up 75% of the resulting data; the prob argument would be like the following: `prob = c(0.25, 0.75)`, where each proportion represents the likelihood of selecting the element of the corresponding level. Table 2.1 depicts a small table that highlights the corresponding elements that the prob argument would reference in another way.

The following code chunk adds this prob argument to the state data attribute. Of note, the data have yet to be generated. After all the data conditions are specified, these will be explored in further detail.

```
sim_arguments <- list(
  formula = achieve ~ 1 + prior_achieve + SES + engagement +
    motivation + state,
  fixed = list(
    prior_achieve = list(var_type = 'continuous'),
    SES = list(var_type = 'factor',
               levels = c('Low', 'Medium', 'High')),
    engagement = list(var_type = 'ordinal'),
    motivation = list(var_type = 'continuous'),
    state = list(var_type = 'factor',
                 levels = c('IA', 'CA'),
                 prob = c(0.25, 0.75))
  )
)
```

2.2.2.1.1 Force Equal Group Size

By default, categorical attributes will be generated based on the abovementioned probability arguments. If the probability arguments are not specified, they will be equal in expectation over the long run or repeated simulations.

However, within a single simulation, the sample size of the categories may not be equal.

It is possible to force equal group sizes when generating categorical attributes. To do this, there is an optional argument for categorical attributes, `force_equal = TRUE`. If the total sample size is divisible by the number of categories for the attribute, then the groups will be sampled until they are of equal size. If the sample size is not divisible by the number of categories for the attribute when the `force_equal = TRUE` is specified, then the groups would be as similar as possible while retaining the overall sample size specified. For example, suppose three categorical groups are being specified (take the SES example from before, with low, medium, and high SES), and the total sample size is 50. When 50 is divided by 3, this would equal 16.67. In this situation, the groups cannot be equal but retain the total sample size of 50; instead, there would be two groups with 17 and one with 16 observations. Across multiple simulations, the group simulated to have 16 observations would change based on how the sampling proceeded.

```
sim_arguments <- list(
  formula = achieve ~ 1 + prior_achieve + SES + engagement +
    motivation + state,
  fixed = list(
    prior_achieve = list(var_type = 'continuous'),
    SES = list(var_type = 'factor',
               levels = c('Low', 'Medium', 'High'),
               force_equal = TRUE), # New force equal code
    engagement = list(var_type = 'ordinal'),
    motivation = list(var_type = 'continuous'),
    state = list(var_type = 'factor',
                 levels = c('IA', 'CA'),
                 prob = c(0.25, 0.75))
  )
)
```

The code above shows how the `force_equal = TRUE` argument can be specified for the SES attribute. In this case, forcing the sample size to be equal while specifying probabilities that are not equal does not match; therefore, only `force_equal` or the `prob` argument should be specified for each attribute.

2.2.2.2 Ordinal Simulation

The `var_type = 'ordinal'` specifies the ordinal simulation of fixed attributes. These types of attributes are those that take on whole integer values but often take on more than a few specific values. One additional argument is needed when specifying an ordinal attribute, `levels`. Below, the ordinal attribute, engagement, is specified as `levels = 1:9` to represent that values 1 through 9

TABLE 2.2 Probability associated with each engagement level using the prob argument.

Engagement	Likelihood
1	0.02
2	0.02
3	0.05
4	0.08
5	0.15
6	0.18
7	0.20
8	0.20
9	0.10

are possible values for this attribute to take. This could represent a rating scale metric where respondents rated their engagement (or maybe their parents) on a scale from 1 to 9.

```
sim_arguments <- list(
  formula = achieve ~ 1 + prior_achieve + SES + engagement +
    motivation + state,
  fixed = list(
    prior_achieve = list(var_type = 'continuous'),
    SES = list(var_type = 'factor',
               levels = c('Low', 'Medium', 'High')),
    engagement = list(var_type = 'ordinal', levels = 1:9),
    motivation = list(var_type = 'continuous'),
    state = list(var_type = 'factor',
                 levels = c('IA', 'CA'),
                 prob = c(0.25, 0.75))
  )
)
```

Similar to the factor attribute specification, it is possible to specify a `prob` argument to give each value a specific likelihood of being selected. The default behavior is that each value is equally likely; therefore, if it would be desired to have middle and upper values of engagement to be more likely to be endorsed, the following `prob` argument could be added: `prob = c(.02, .02, .05, .08, .15, .18, 0.2, 0.2, 0.1)`. Each element of the prob argument is associated with a level of the engagement attribute from smallest to largest.

TABLE 2.3 Statistical distributions and the associated random distribution functions available within R by default.

Distribution	Function
Beta	'rbeta()'
Binomial	'rbinom()'
Cauchy	'rcauchy()'
Chi-Square	'rchisq()'
Exponential	'rexp()'
F	'rf()'
Gamma	'rgamma()'
Geometric	'rgeom()'
Hypergeometric	'rhyper()'
Logistic	'rlogis()'
Log Normal	'rlnorm()'
Negative Binomial	'rnbinom()'
Normal	'rnorm()'
Poisson	'rpois()'
Student t	'rt()'
Studentized Range	'rtukey()'
Uniform	'runif()'
Weibull	'rweibull()'
Wilcoxon Rank Sum Statistic	'rwilcox()'
Wilcoxon Signed Rank Statistic	'rsignrank()'

2.2.2.3 Continuous Simulation

Continuous data attributes are simulated using the `var_type = 'continuous'` argument. The only other primary argument for simulation continuous data attributes is a random distribution function through the `dist` argument. One common example of a random distribution function is the `rnorm()` function, which would generate random draws from the normal distribution. Any valid random distribution function is possible. Table 2.3 gives a list of the name and the function to show random distribution functions available within R by default. More may be available from other specific packages and could be specified if that package is first loaded.

Depending on which random distribution function is specified, additional optional arguments for that distribution may be needed. Two examples from two different distributions are given below to generate a continuous normal attribute and a continuous attribute that is negatively skewed. First, the prior achievement attribute is specified as a normal attribute with the argument, `dist = 'rnorm'`. For the normal distribution, other arguments are not needed, and the distribution would be simulated from the standard normal distribution with a mean of 0 and a standard deviation of 1. If users wish to alter these, the `mean` and `sd` arguments can be specified to alter the mean and standard deviation of the attribute. This is done below to generate the prior achievement attribute with a mean of 100 and a standard deviation of 10.

To generate a negatively skewed distribution, the Beta distribution can be used with the argument, `dist = 'rbeta'`. For the `rbeta()` random generation function, shape arguments must be specified with `shape1` and `shape2`. To get a positively skewed distribution, the `shape2` argument should be larger than the `shape1` argument. The opposite should be specified for a negatively skewed distribution; specifically, the `shape1` argument should be larger than the `shape2` argument. An optional non-centrality parameter (`ncp`) argument can be specified here too, but this is left to the reader.

```
sim_arguments <- list(
  formula = achieve ~ 1 + prior_achieve + SES + engagement +
    motivation + state,
  fixed = list(
    prior_achieve = list(var_type = 'continuous', dist = 'rnorm',
                         mean = 100, sd = 10),
    SES = list(var_type = 'factor',
               levels = c('Low', 'Medium', 'High')),
    engagement = list(var_type = 'ordinal', levels = 1:9),
    motivation = list(var_type = 'continuous', dist = 'rbeta',
                      shape1 = 5, shape2 = 1),
    state = list(var_type = 'factor',
                 levels = c('IA', 'CA'),
                 prob = c(0.25, 0.75))
  )
)
```

2.2.2.3.1 Floor or Ceiling Effects

It is also possible to simulate floor or ceiling effects for continuous attributes. This may be of interest in a situation when data are simulated that would be distributed in a specific way, but through the measurement process, the scores are truncated by a floor or ceiling value. One common example of the presence of a floor effect is for air quality, where through the measurement process, values can not go below 0. Ceiling effects could be possible for very easy exams where many individuals may achieve a perfect score so that the ceiling effect would impose this constraint.

To impose floor or ceiling effects, there are optional arguments when specifying a continuous attribute named `floor` and `ceiling` that would impose floor or ceiling constraints, respectively. It is possible to use both for the same attribute as well. Below is an example of specifying these for two continuous attributes.

```
sim_arguments <- list(
  formula = y ~ 1 + age + exam,
  fixed = list(age = list(var_type = 'continuous',
                          mean = 6, sd = 1,
                          floor = 5),
               exam = list(var_type = 'continuous',
                           mean = 90, sd = 6,
                           ceiling = 100))
)
```

2.2.2.4 Show Fixed Data

Now that the fixed data have been fully specified, the fixed data can be simulated using the `simulate_fixed()` function. The only piece left to specify is the number of observations to simulate. This is added to the simulation arguments via the named element of the list as `sample_size`. Below it is specified as `sample_size = 1000` to simulate 1,000 rows of data. The first six rows are shown.

```
sim_arguments <- list(
  formula = achieve ~ 1 + prior_achieve + SES + engagement +
    motivation + state,
  fixed = list(
    prior_achieve = list(var_type = 'continuous', dist = 'rnorm',
                         mean = 100, sd = 10),
    SES = list(var_type = 'factor',
               levels = c('Low', 'Medium', 'High')),
    engagement = list(var_type = 'ordinal', levels = 1:9),
    motivation = list(var_type = 'continuous', dist = 'rbeta',
                      shape1 = 5, shape2 = 1),
    state = list(var_type = 'factor',
                 levels = c('IA', 'CA'),
                 prob = c(0.25, 0.75))
  ),
  sample_size = 1000
)

simulate_fixed(data = NULL, sim_arguments)
```

The following output shows that several columns of data are generated. A few notes should be made about the specifics of the data generated as more data is returned than what is processed. This is for added transparency, but is helpful to understand later to specify the correct number of other dimensions—first, the `X.Intercept.` term is specified and corresponds with the `1` in the model formula. The prior achievement, engagement, and motivation attributes are returned once and represent data attributes as simulated.

```
##   X.Intercept. prior_achieve SES_1 SES_2 engagement
## 1            1        102.32     1     0          2
## 2            1         90.39     0     0          2
## 3            1        105.17     0     0          3
## 4            1        117.38     0     1          4
## 5            1         88.76     0     0          8
## 6            1         97.74     0     1          6
##   motivation state_1    SES state level1_id
## 1     0.4925       1 Medium    CA         1
## 2     0.9477       1    Low    CA         2
## 3     0.9792       0    Low    IA         3
## 4     0.7808       0   High    IA         4
## 5     0.9969       1    Low    CA         5
## 6     0.9413       0   High    IA         6
```

The elements in the output that show up more than once include SES and state elements. More generally, within the **simglm** package, more than one column will show up for data that are simulated as factor attributes. For these attributes, the default behavior is to create dummy or indicator codes for those data attributes. The data shows these attributes under the column names, `SES_1`, `SES_2`, and `state_1`. These are common ways to represent the factor attributes in a linear modeling framework, where the number of factor categories minus one additional attributes are added for each factor simulated.

To be more specific, the `SES_1` attribute represents those with "Low" SES, the `SES_2` attribute represents those with "Medium" SES, and the `state_1` attribute represents those from "IA". The default behavior is for those closer to "A" to be a reference group and aren't given a specific attribute. More details on dummy or indicator coding can be found at Fox (2015). It is important to take some time to understand how these elements are specified in the linear modeling framework, which will impact how many β weights need to be specified—more details on this to come.

2.2.3 Define β Weights

The β weights in linear regression are a way to represent the relationship between each attribute (i.e., the X terms in Equation (2.1)) and the outcome. In general, these are slope terms where the β weight is the change in the outcome for a one-unit change in the associated attribute.

TABLE 2.4 Regression coefficient (β weight) associated with each term in the regression formula.

Term	Coefficient
Intercept	2.00
Prior Achievement	0.25
SES_1 (Medium)	0.50
SES_2 (High)	1.20
Engagement	0.10
Motivation	0.50
State_1	−1.50

To determine the number of β weights needed to be specified in order to simulate the data, there will be one associated with each continuous or ordinal fixed effect. For categorical attributes, the default behavior is to create a dummy or indicator variable, and there will need to be the number of categories minus one for each categorical attribute. These are specified as a named numeric vector within the simulation arguments, named `reg_weights`. Below is an example of specifying these for the example used already.

The regression coefficients (i.e., β weights) are specified in the order of the model formula. To clarify, Table 2.4 shows the regression coefficients and the term they are associated with—specific care to associate the correct terms with the regression weights so that the proper simulation is specified. The most challenging part is often making sure that enough regression coefficients are specified for categorical terms, as it only shows up once in the formula. However, if more than two groups are simulated, more than one regression coefficient must be specified for the data simulation to process.

2.2.4 Define and Specify Random Error

The random error represents variation in the outcome that is noise or unexplained rather than signal or explained variation. In Equation (2.1) the ϵ is the notation for the random error. Another way to think about the random error is the deviation of each individual from the predicted or model-implied regression line.

Within the **simglm** package, the default error is set to have a variance of one and come from a normal distribution. If this is desired, no code is needed. However, the slightly more verbose code can aid in interpreting what is being simulated. Therefore, it is recommended to add a named element, named 'error,' to the simulation arguments. Below, the error variance is added to have a variance of 5. As no distribution was specified, it would be simulated from the random normal distribution (i.e., `rnorm()`).

```
sim_arguments <- list(
  formula = achieve ~ 1 + prior_achieve + SES + engagement +
    motivation + state,
  fixed = list(
    prior_achieve = list(var_type = 'continuous', dist = 'rnorm',
                         mean = 100, sd = 10),
    SES = list(var_type = 'factor',
               levels = c('Low', 'Medium', 'High')),
    engagement = list(var_type = 'ordinal', levels = 1:9),
    motivation = list(var_type = 'continuous', dist = 'rbeta',
                      shape1 = 5, shape2 = 1),
    state = list(var_type = 'factor',
                 levels = c('IA', 'CA'),
                 prob = c(0.25, 0.75))
  ),
  sample_size = 1000,
  reg_weights = c(2, .25, 1.2, 0.5, .10, .5, -1.5),
  error = list(variance = 5)
)
```

It is possible to have the error distribution be something other than a normal distribution, however. For example, suppose the errors were desired to have a slightly heavier tail; a t-distribution could be used for this. Similar to simulating continuous attributes, the `dist` argument can be specified within the named `error` element of the simulation arguments. This is shown below using the t-distribution random number generator, `rt()`. The `rt()` function also needs the degrees of freedom to be specified, and in this case, this was specified as `df = 4` to represent a t-distribution with 4 degrees of freedom.

```
sim_arguments <- list(
  formula = achieve ~ 1 + prior_achieve + SES + engagement +
    motivation + state,
  fixed = list(
    prior_achieve = list(var_type = 'continuous', dist = 'rnorm',
                         mean = 100, sd = 10),
    SES = list(var_type = 'factor',
               levels = c('Low', 'Medium', 'High')),
    engagement = list(var_type = 'ordinal', levels = 1:9),
    motivation = list(var_type = 'continuous', dist = 'rbeta',
                      shape1 = 5, shape2 = 1),
    state = list(var_type = 'factor',
                 levels = c('IA', 'CA'),
                 prob = c(0.25, 0.75))
```

TABLE 2.5 First six rows of the simulated random error.

error	level1_id
1.7096	1
0.2239	2
2.1878	3
4.1312	4
0.9257	5
−3.3988	6

```
  ),
  sample_size = 1000,
  reg_weights = c(2, .25, 1.2, 0.5, .10, .5, -1.5),
  error = list(variance = 5, dist = 'rt', df = 4)
)

simulate_error(data = NULL, sim_arguments)
```

Lastly, the `simulate_error()` function can be used to generate the error. The first six rows of the generated error are shown in Table 2.5. Additional optional arguments for the random error will be explored later, including simulated variance heterogeneity.

2.2.5 Generate Outcome

Now that all the essential elements of a linear regression, as shown in Equation (2.1), the outcome or Y attribute can be generated as a function of the fixed effects, regression coefficients, and the random error. The generation of the outcome is done with the `generate_response()` function.

The steps can be combined, and the three functions needed, `simulate_fixed()`, `simulate_error()`, and `generate_response()`, can be chained together with the R pipe operator, `|>`. Note that the simulation arguments are specified as before and are not repeated here. Finally, a seed is specified to ensure the generated outcomes are reproducible for this code section.

```
set.seed(2022)

simulate_fixed(data = NULL, sim_arguments) |>
  simulate_error(sim_arguments) |>
  generate_response(sim_arguments)
```

TABLE 2.6 First six rows of the simulated linear regression model. The prior fixed components are omitted from this table.

level1_id	error	fixed_outcome	random_effects	achieve
1	0.4333	28.84	0	29.27
2	1.6264	24.47	0	26.10
3	−2.0169	25.53	0	23.51
4	0.1512	22.75	0	22.90
5	−1.5580	27.36	0	25.80
6	−5.4040	20.23	0	14.83

The first function in the chain could be either `simulated_fixed()` or `simulate_error()`, but the first one specified needs to tell the function that there is no data to build off of with the `data = NULL` argument. After that, the second argument for each function in **simglm** is the simulation arguments, so those are specified for each function in the chain. With the pipe, the data simulated from the preceding step is inputted as the first argument. Table 2.6 shows the first six rows of the simulated data.

Notice that this data now includes all of the pieces included before but now includes a few new columns of data, including `fixed_outcome`, `random_effects`, and `achieve`. These would represent the multiplication of $X\beta$, a placeholder column for random effects discussed in more detail in Chapter 3, and the achievement outcome (i.e., the Y from Equation (2.1)).

A linear regression could then be fitted to these data to estimate the parameters of interest. Table 2.7 shows the estimated coefficients for this model. Notice that the estimated coefficients are similar, but not exact, to the specified regression weights in the simulation arguments. This would be expected due to the random nature of the simulation process. For a single simulated data, these parameters would tend to get closer to the specified values if the sample size

TABLE 2.7 Estimated regression coefficients from simulated data.

term	estimate	std.error	statistic	p.value
(Intercept)	0.4334	1.0963	0.3954	0.6927
prior_achieve	0.2668	0.0091	29.2755	0.0000
SESMedium	1.3450	0.2224	6.0487	0.0000
SESHigh	0.6913	0.2231	3.0987	0.0020
engagement	0.1044	0.0357	2.9260	0.0035
motivation	0.5506	0.6641	0.8290	0.4073
stateCA	−1.8552	0.2125	−8.7305	0.0000

was included, but the estimated parameters would center around the specified values if this process were replicated many times with the current sample size.

```
set.seed(2022)

sim_data <- simulate_fixed(data = NULL, sim_arguments) |>
  simulate_error(sim_arguments) |>
  generate_response(sim_arguments)

lm(achieve ~ 1 + prior_achieve + SES + engagement + motivation + state,
   data = sim_data)
```

3

Simulating General Linear Mixed Models

3.1 Chapter Overview

This chapter extends simulation methods to general linear mixed models, focusing on data with hierarchical or clustered structures. The chapter discusses how multilevel data arise in applied settings such as education, psychology, and health research, and how ignoring nesting can distort inference. Simulation strategies for random intercept and random slope models are presented, including the specification of variance components, intraclass correlations, and cross-level interactions. The chapter emphasizes how choices about cluster sizes, number of clusters, and variance structures affect estimation and power. Examples illustrate the impact of clustering on standard errors and treatment effect detection. By simulating mixed models directly, applied researchers are shown how to explore design questions that involve complex dependency structures and how multilevel assumptions influence results.

General linear mixed models (LMM) extend the LM model discussed in Chapter 2 by adding error-like terms called random effects. These random effects are commonly group-specific deviations from the average line of best fit. The LMM model is often used in situations with data nesting, hierarchies, or other situations where the data at the unit of analysis are not independent. The LMM can be written in general matrix form as:

$$\mathbf{Y} = \mathbf{X}\beta + \mathbf{Z}b_j\epsilon \tag{3.1}$$

where $\mathbf{Y}$ is a vector representing the outcome attribute, $\mathbf{X}$ is a matrix of data attributes representing fixed effects, β is a vector of regression coefficients, $\mathbf{Z}$ is a matrix of data attributes for the random effects, b_j are random effects, and ϵ represents random error. It is beyond the scope of this book to give an entirely in-depth introduction and description of these models but to understand better how to simulate data from a model like this and how these correspond to the functions used in the **simglm** package, a brief description of random effect terms are discussed. Those interested in a more in-depth introduction to the LMM model are directed to (Fitzmaurice et al., 2012; Goldstein, 2011; Raudenbush and Bryk, 2002). The LMM model has many names within different research fields, sometimes referred to as multilevel

DOI: 10.1201/9781003177876-3

models (Goldstein, 2011), hierarchical models (Raudenbush and Bryk, 2002), or covariance components models (Dempster et al., 1981). This can make it difficult to follow these methods, but the LMM name will be used throughout this book.

3.2 Short Description of Random Effects

Random effects have a variety of definitions within the statistical literature and often depend on which field the study arises in. This book defines random effects from the LMM literature (Fitzmaurice et al., 2012), where random effects are defined as subject-specific effects unique to those units. This is best articulated with a few examples.

Imagine that data from multiple measurement occasions were collected from a group of individuals. In this situation, the measurement occasions would be nested within the individuals. There are multiple ways to model this situation to account for the correlated observations from measuring the same individuals at different times. One such way is with the LMM. In the LMM framework, the random effects would represent individuals' specific effects that are unique to each individual. Suppose in this example that we are measuring physical activity every week over the course of a summer. The random effects in this situation may reflect whether an individual tends to exercise more or less than the average of those included in the sample. A second random effect could also be included, which could reflect whether an individual changes their physical activity more or less than the average change in the sample. The average physical activity and change in physical activity over the course of the study reflect the fixed effects in this framework. Inside the LMM framework, the fixed effects could also be thought of as population effects.

Another example to generalize the idea of random effects in the LMM framework outside of longitudinal data. Imagine collecting data on melanoma cancer treatment success rates of hospitals across different countries. In this situation, the hospitals would be nested within countries. As with the previous example, there would be multiple ways to model this data, but one such way is within an LMM framework. In this framework, the fixed effects would represent the average melanoma cancer treatment success rate for all hospitals. The random effect would represent country-specific effects, namely, whether that country has an above- or below-average melanoma cancer treatment success rate across all hospitals. A positive random effect would represent a country that has better treatment outcomes. In contrast, a negative random effect would represent a country that has worse treatment outcomes compared to the average.

3.3 Simulating the General Linear Mixed Model

Many of the steps for simulating the general linear mixed model are the same as those for the general linear model discussed in Chapter 2. This chapter will build off the example already discussed from Chapter 2, focusing on elements that differ when simulating data from the LMM instead of the LM.

The steps to simulating from the LMM, in the simplest case, can be broken down into about seven steps. These steps are written in words below with some indication about what function when using the **simglm** package; each will be discussed in more detail:

1. Specify the terms for the model; for example, how many fixed terms are there?
2. Define and specify the simulation terms for the fixed attributes.
3. Define and specify the simulation terms for the random effects.
4. Define and specify the simulation terms for the random error.
5. Define the relationship between the fixed attributes ($\mathbf{X}$) and the outcome attribute ($\mathbf{Y}$), that is the β weights.
6. Based on steps 2 through 5, simulate
   - the fixed terms (e.g., using `simulate_fixed()`)
   - the random effects (e.g., using `simulate_randomeffect()`)
   - the random error (e.g., using `simulate_error()`)
7. Generate the outcome based on steps 2 through 5 (e.g., using `generate_response()`).

The primary difference when simulating data with the LMM model instead of the LM model is step 3. Step 3 defines the population characteristics for the random effects. The simulated data from Chapter 2 had a column in the simulated data for the LM named `random_effects`. This is a placeholder for the current LMM. In Chapter 2, the random effect terms were assumed to be zero; thus, they dropped from the equation.

3.3.1 Define and Specify Random Effects

Recall in Chapter 2 that the simulated example was attempting to predict student achievement with other data attributes. Often, in student achievement data, the data are nested. This occurs because students are more similar within a specific group (i.e., classroom, school, etc.) compared to between those same groups. A specific example with the student achievement data is that students within a school are more similar than students from a different school. This occurs within a single city, even as schools are often neighborhood-based, and neighborhoods can vary by socioeconomic status, demographics, or other features.

Using the LMM can account for the data's nesting by including random effects. In the **simglm** package, the definition and specification of random effects is done through the `randomeffect` named simulation arguments. The package also adopts syntax that is similar to the **lme4** R package (Bates et al., 2015) for including random effects in the model formula.

Below the, the formula and random effect arguments are specified to create a random intercept for each school in the data. This example is intended to build on the prior example in Chapter 2, but all other arguments were removed to focus on what is different. In the formula argument, the following element was added to specify the random intercept, `(1 | school)`. The syntax can be read as `(random effects | nesting ID)` within the parentheses. Being more verbose, the random effects are the random effects defined to the left of the vertical bar (i.e., `|`), and the ID that is associated with those random effects is to the right of the vertical bar. When simulating the data, the name of the ID column for the random effects is specified to the right of the vertical bar.

```
sim_arguments <- list(
  formula = achieve ~ 1 + prior_achieve + SES + engagement +
    motivation + state + (1 | school),
  randomeffect = list(
    int_school = list(variance = 3, var_level = 2)
  ),
  sample_size = list(level1 = 10, level2 = 8)
)
```

The `randomeffect` specification is similar to the fixed effect specification in that a nested list is used. Within the list, named arguments for the random effects are specified. In the current example, only one random effect is specified in the formula. Therefore, a single named element within the `randomeffect` simulation argument is needed. In this case, the name `int_school` will be the name for that random effect in the generated data. Within this, two required elements are defined. First is the variance of that random effect. Similar to the random error from Chapter 2, the random effects are specified with a variance term. Secondly, the level that this random effect is specified on is determined with the argument `var_level`. This argument could take on values of 2 or 3, representing a nested attribute that is at level 2 or level 3, respectively.

Finally, as the data are now nested, the sample size for these levels now needs to be specified. The simplest specification assumes that the data are balanced; an example is shown above. The sample size is now a list with two named elements, `level1` and `level2`. A single numeric value is specified for both. In the context of the current example, the `level2 = 8` means that there would be

TABLE 3.1 First six rows of the simulated random effect.

int_school	level1_id	school
−1.25	1	1
−1.25	2	1
−1.25	3	1
−1.25	4	1
−1.25	5	1
−1.25	6	1

eight schools, and `level1 = 10` means that within each school, there would be ten students. The total sample size (i.e., the number of rows in the simulated data) could be found by multiplying the two values, $8 * 10 = 80$.

3.3.1.1 Simulate Random Effects

To perform the random effect simulation, the `simulate_randomeffect()` function is used. This function can be used within a chain or can be used as a standalone function. For example, to simulate just the random effect from the prior simulation argument, the following code will simulate that.

```
simulate_randomeffect(data = NULL, sim_arguments)
```

Notice from the output that there are a total of 80 columns in the data. Two of those columns are ID columns that are used to identify which school (level two unit, named "school") the observation belongs to, plus which individual within each school (named "level1_id"). These two ID columns uniquely identify an observation.

The third column (the first in the output), named "int_school," is the name for the random effect specified inside the random effect element of the list inside `sim_arguments`. This column represents the school-specific deviation in achievement. Therefore, those that are positive indicate, on average, that the school performs better than an average school. The opposite is true for those that have a negative random effect. Notice also that these are repeated for each level one ID within a specific school. This indicates that these are the school effects and are the same for each individual. There are a total of eight unique school effects in these simulated data as the level two sample size was specified as eight.

The sample size can be shown using the `table()` or `count()` function. The number of unique values for the school column can also show the level 2 sample size, 8. This can be shown more detail with the `table()` function.

TABLE 3.2 Sample size within each unique school ID value for the simulated random effects.

Var1	Freq
1	10
2	10
3	10
4	10
5	10
6	10
7	10
8	10

```
table(
  simulate_randomeffect(data = NULL, sim_arguments)$school
)
```

3.3.1.2 Unbalanced Sample Size

If the balanced sample size is not commonly collected with applied or real-world data, then unbalanced sample sizes can be simulated. There are a few ways to do this, but generally, a numeric vector of sample sizes can be added to the "sample_size" element of the simulation arguments. This vector could be specified manually using the `c()` function to create a numeric vector or the unbalanced sample sizes can be generated in a random fashion with `runif()` or `sample()`. The primary consideration to ensure is that the length of the numeric vector needs to match the sample size for the level two sample size. To state it another way, the unbalanced sample size here represents a different number of observations within each level two unit. For example, the number of students in a single school differs from the other schools.

The code below uses the `runif()` function to generate eight random values from the uniform distribution. The min and max arguments are used to specify that the uniform distribution should start at 4 and range up to 20. Finally, as these are sample sizes, whole integer values are needed; therefore, the `round()` function is used to round to a whole numeric number.

```
set.seed(1)

level1_ss <- round(runif(8, min = 4, max = 20), 0)
level1_ss
```

```
## [1]  8 10 13 19  7 18 19 15
```

Once this has been specified, this can be specified directly within the "sample_size" simulation arguments. Notice that the named object, "level1_ss", is used in place of the single numeric value that was specified before. Once the unbalanced sample sizes are added, the `simulate_randomeffect()` function can be used as before to generate the random effects.

```
sim_arguments <- list(
  formula = achieve ~ 1 + prior_achieve + SES + engagement +
    motivation + state + (1 | school),
  randomeffect = list(
    int_school = list(variance = 3, var_level = 2)
  ),
  sample_size = list(level1 = level1_ss,
                     level2 = 8)
)
```

As shown above, the `table()`, (or `count()`) function can be used to generate the number of observations within each level two unit (i.e., schools). Table 3.3 shows the specified sample sizes. These are shown to be the same as those shown above when creating the `level1_ss` object. It also is important to note that the unbalanced sample sizes are used by position. Therefore, the first level two unit is paired with the first element of the unbalanced sample size vector. This is important to consider when specifying the sample sizes manually.

```
table(
  simulate_randomeffect(data = NULL, sim_arguments)$school
)
```

TABLE 3.3 Unbalanced sample size within each unique school ID value for the simulated random effects.

Var1	Freq
1	8
2	10
3	13
4	19
5	7
6	18
7	19
8	15

3.3.2 Add Random Slope

Adding a random slope is similar to that of fixed effects. Slopes are identified by the name of an attribute within the `()` part of the formula. Usually, in the case of a random slope, the random slope is an attribute that was already specified as part of the fixed portion of the model. The example below will add a random slope for the term "motivation." This was done with the `(1 + motivation | school)` portion added to the model formula.

In addition to the part added to the model formula, the population parameters (i.e., the "truth") for the random slope need to be specified. This is done within the "randomeffect" named element of the simulation arguments and is shown below with the "mot_school" term below. The `variance` argument and `var_level` arguments are needed to be specified. The variance is how much variation one would expect in the random slope term, here defined as 0.5. In general, with real-world data, random slopes show less variation than random intercepts. The `var_level` argument says which level the random slope should be specified at, in this case, level two or school level random attribute. Finally, two additional optional arguments change the distribution from which these are generated. This was done by specifying `dist = 'rt'`, which is a t-distribution. The `rt()` function needs the degrees of freedom defined as well, and this was included with `df = 5`.

```
sim_arguments <- list(
  formula = achieve ~ 1 + prior_achieve + SES + engagement +
    motivation + state + (1 + motivation | school),
  randomeffect = list(
    int_school = list(variance = 3, var_level = 2),
    mot_school = list(variance = 0.5, var_level = 2,
                      dist = 'rt', df = 5)
  ),
  sample_size = list(level1 = level1_ss, level2 = 8)
)
```

The `simulate_randomeffect()` function is still used to generate these random effects as before. The primary difference in the first six rows shown in Table 3.4 is the additional column for the new random effect, "mot_school." This was the same name as specified within the simulation arguments. These random effects are interpreted at the school level, just like the random intercept; therefore, each level two unit has a unique random slope. As such, these terms are interpreted as the average deviation in the motivation effect for each school.

```
simulate_randomeffect(data = NULL, sim_arguments)
```

TABLE 3.4 First six rows of the simulated random effect with a random slope.

int_school	mot_school	level1_id	school
−1.076	0.5691	1	1
−1.076	0.5691	2	1
−1.076	0.5691	3	1
−1.076	0.5691	4	1
−1.076	0.5691	5	1
−1.076	0.5691	6	1

3.4 A Complete Example with Random Effects

The example below combines the entire pipeline, adding the same simulation arguments from Chapter 2. Instead of being the linear regression from Chapter 2, the data are now multilevel with school effects.

```
sim_arguments <- list(
  formula = achieve ~ 1 + prior_achieve + SES + engagement +
    motivation + state + (1 + motivation | school),
  randomeffect = list(
    int_school = list(variance = 3, var_level = 2),
    mot_school = list(variance = 0.5, var_level = 2,
                      dist = 'rt', df = 5)
  ),
  sample_size = list(level1 = level1_ss,
                     level2 = 8),
  fixed = list(
    prior_achieve = list(var_type = 'continuous', dist = 'rnorm',
                         mean = 100, sd = 10),
    SES = list(var_type = 'factor',
               levels = c('Low', 'Medium', 'High')),
    engagement = list(var_type = 'ordinal', levels = 1:9),
    motivation = list(var_type = 'continuous', dist = 'rbeta',
                      shape1 = 5, shape2 = 1),
    state = list(var_type = 'factor',
                 levels = c('IA', 'CA'),
                 prob = c(0.25, 0.75))
  ),
  reg_weights = c(2, .25, 0.5, 1.2, .10, .5, -1.5),
```

```
  error = list(variance = 5, dist = 'rt', df = 4)
)
```

The data simulation process proceeds as follows, 1. simulate fixed effects 2. simulate random effects 3. simulate random error 4. generate response.

Note that the simulation process does not necessarily need to proceed in this way; for example, random effects or random errors could be generated first. This matches procedurally how the model was built and works from left to right based on how the models are typically written out. The resulting data are shown in the following output.

```
set.seed(2022)

simulate_fixed(data = NULL, sim_arguments) |>
  simulate_randomeffect(sim_arguments) |>
  simulate_error(sim_arguments) |>
  generate_response(sim_arguments)
```

```
##   X.Intercept. prior_achieve SES_1 SES_2 engagement
## 1            1        109.00     0     1          9
## 2            1         88.27     1     0          4
## 3            1         91.03     0     0          7
##   motivation state_1    SES state level1_id school
## 1     0.7215       1   High    CA         1      1
## 2     0.8526       1 Medium    CA         2      1
## 3     0.8974       0    Low    IA         3      1
##   int_school mot_school    error fixed_outcome
## 1    0.03985  -0.008553 -2.63378         30.21
## 2    0.03985  -0.008553 -1.37588         23.89
## 3    0.03985  -0.008553  0.05771         25.91
##   random_effects achieve
## 1        0.03367   27.61
## 2        0.03255   22.55
## 3        0.03217   25.99
```

3.5 Specialized Case: Longitudinal Data

Longitudinal data are a special case of data nesting, multilevel, or mixed models. In the simplest case, the repeated observations are collected within an individual. Therefore, knowing something about a prior observation gives information about subsequent observations. For example, if a person's anxiety level (i.e., collected via an anxiety screening assessment) is measured in September of a year, then again in December, these two observations would be correlated. This correlation can and should be modeled, and linear mixed models are one way to account for this correlation.

The model code below specifies a structure for repeated anxiety measurements across five-time points within 20 people (look at the sample size arguments to confirm these values). The default structure for generating the time attribute from the model formula is to start at 0, proceed in increments of 1, and go to 1 less than the level 1 sample size. Since the level one sample size is 5, the time attribute would be 0, 1, 2, 3, 4. This can be modified and will be discussed in the next section, but this is the default behavior when the `var_type = 'time'` is specified with no other optional arguments specified.

The random effect simulation arguments are specified the same as before. This example has a random intercept and a random slope for time. This structure would show that the starting point for anxiety of individuals differs (random intercept), and the linear trend would also vary for each individual (random slope).

Finally, two other elements to note here. First, the random error is not directly specified in the simulation arguments. Therefore, this would be generated from a standard normal distribution (i.e., mean of zero, variance of one). Secondly, notice how the "treat" attribute has an optional argument of `var_level = 2`. This would generate this attribute at the individual level. Procedurally, this could be a situation when individuals are randomly assigned to treatment or control at the start of the study and do not switch treatments over time. If the `var_level = 1` were specified instead, this would be a situation when the individuals could switch between the treatment or control at each measurement occasion, a situation that is unlikely to be done in practice.

```
set.seed(321)

sim_arguments <- list(
  formula = anxiety ~ 1 + time + treat + (1 + time| id),
  fixed = list(time = list(var_type = 'time'),
               treat = list(var_type = 'factor',
                            levels = c('Control', 'Treatment'),
                            var_level = 2)),
  randomeffect = list(int_id = list(variance = 8, var_level = 2),
                      time_id = list(variance = 3, var_level = 2)),
  sample_size = list(level1 = 5, level2 = 20),
  reg_weights = c(0.5, 0.25, 0.3)
)
```

The code below simulates the longitudinal data. The pathway is the same as in previous examples in this chapter. The output shows that the simulated longitudinal data structure is similar; the most notable new feature is the "time" attribute that starts at 0 and increments by 1. This is the default specification for time using the **simglm** package.

```
set.seed(2022)

simulate_fixed(data = NULL, sim_arguments) |>
  simulate_randomeffect(sim_arguments) |>
  simulate_error(sim_arguments) |>
  generate_response(sim_arguments)
```

```
##   X.Intercept. time treat_1     treat level1_id id
## 1            1    0       1 Treatment         1  1
## 2            1    1       1 Treatment         2  1
## 3            1    2       1 Treatment         3  1
## 4            1    3       1 Treatment         4  1
## 5            1    4       1 Treatment         5  1
## 6            1    0       0   Control         1  2
##    int_id time_id  error fixed_outcome random_effects
## 1  2.8459   0.204 1.6932          0.80         2.8459
## 2  2.8459   0.204 0.9958          1.05         3.0499
## 3  2.8459   0.204 0.1868          1.30         3.2539
## 4  2.8459   0.204 1.2383          1.55         3.4579
## 5  2.8459   0.204 0.3094          1.80         3.6619
## 6 -0.5237   1.440 0.6357          0.50        -0.5237
##   anxiety
## 1   5.339
## 2   5.096
## 3   4.741
## 4   6.246
## 5   5.771
## 6   0.612
```

3.5.1 Manual Time Specification

The default time specification can be helpful for a quick analysis. However, if a more specific time collection wants to be specified, it can be with an optional time simulation argument named `time_levels`. Suppose, for example, data are collected for a year every quarter (i.e., every three months) with a baseline data collection. The code example below is updated to include the `time_levels` optional argument to specify this. The only restriction with the `time_levels` argument is that the vector needs to be the same length as the level one sample size.

TABLE 3.5 Unique time points based on the manual time specification.

time	n
0	20
3	20
6	20
9	20
12	20

```
set.seed(321)

sim_arguments <- list(
  formula = anxiety ~ 1 + time + treat + (1 + time| id),
  fixed = list(time = list(var_type = 'time',
                           time_levels = c(0, 3, 6, 9, 12)),
             treat = list(var_type = 'factor',
                          levels = c('Treatment', 'Control'),
                          var_level = 2)),
  randomeffect = list(int_id = list(variance = 8, var_level = 2),
                    time_id = list(variance = 3, var_level = 2)),
  sample_size = list(level1 = 5, level2 = 20)
)
```

Instead of generating all of the data as before, the fixed effects will be generated with `simulate_fixed()`, and the unique value for the time attribute will be shown. Table 3.5 shows the unique time values for the fixed effect of time. These mimic those that were generated with the optional `time_values` argument.

3.6 Add a Third Level

Mixed models can contain more than one level of nesting. For example, instead of having students nested solely within schools, another natural level could be neighborhoods. It is common for students to go to schools that are within their neighborhood, at least in the United States. Students within a particular neighborhood may also be more similar than across neighborhoods, adding another potential portion of variation that this nesting structure can explain. This structure can be simulated with **simglm** as well. The syntax is also similar to that of the **lme4** package. An additional level of nesting can be added with a second `()` in the model formula. Any random components that are defined in that second set of parentheses would have matching specifications in the "randomeffect" named element of the simulation arguments.

The example code below includes a random intercept for each unique neighborhood with the formula syntax, `(1 | neighborhood)`. The population parameters for this random effect are defined in the "int_neighborhood" section of the randomeffect simulation arguments. The variance was defined to equal one and the `var_level = 3`, which species this attribute should generate at level 3: a unique value for each unique level 3 unit or neighborhood. Finally, one more specification for the sample size was given in the "sample_size" named argument. Since an additional level of nesting was added, the sample size and number of units associated with them need to be specified. Here, four unique neighborhoods are generated.

```
sim_arguments <- list(
  formula = achieve ~ 1 + prior_achieve + SES + engagement +
    motivation + state + (1 + motivation | school) + (1 | neighborhood),
  randomeffect = list(
    int_school = list(variance = 3, var_level = 2),
    mot_school = list(variance = 0.5, var_level = 2,
                      dist = 'rt', df = 5),
    int_neighborhood = list(variance = 1, var_level = 3)
  ),
  sample_size = list(level1 = 8,
                     level2 = 8,
                     level3 = 4)
)
```

Table 3.6 shows the resulting random effects simulated with the `simulate_randomeffect()` function. Two additional columns were added compared to the previous example in Table 3.4, one for the neighborhood IDs and another representing the random neighborhood intercepts, named "int_neighborhood."

TABLE 3.6 First six rows of the simulated random effect with a third level of nesting.

int_school	mot_school	int_neighborhood	level1_id	school	neighborhood
0.6022	0.1822	−0.2776	1	1	1
0.6022	0.1822	−0.2776	2	1	1
0.6022	0.1822	−0.2776	3	1	1
0.6022	0.1822	−0.2776	4	1	1
0.6022	0.1822	−0.2776	5	1	1
0.6022	0.1822	−0.2776	6	1	1

```
simulate_randomeffect(data = NULL, sim_arguments)
```

3.6.1 Unbalanced Level One and Two Data Simulation

When specifying three-level data, sample sizes often differ across different data groups. For example, suppose there is a situation where the nesting structure is individuals nested in classrooms nested in schools. Schools are naturally of different sizes, and if all classrooms from a specific grade are included in the sampling plan, there would be expected to be different classrooms within each school. Similarly, the number of individuals within a classroom will vary within and across schools. For example, perhaps one school has larger classroom sizes with more aides, and another has smaller classes but no classroom aids. Furthermore, differences between rural and urban schools are possible. For this reason, building this unbalanced sample size structure into the data simulation process is often of interest. This aids in having more realistic data, which is often desirable.

There are two broad categories for which this can be specified in the **simglm** package. 1. Using a random generator, for example, `sample()` or `runif()` in tandem with `round(0)`. 2. Specify the sample sizes manually with a numeric vector. The part that can be tricky is ensuring that the vectors are the correct size based on sample sizes specified for other levels. For this reason, using random number generators can make it easier to specify how many numbers should be generated.

3.6.1.1 Unbalanced Level 2 Simulation

Imagine collecting data on business culture via a survey. A survey was sent to all of the employees at eight different businesses. Before sending out the survey, study planning via simulation was desired. This would be an important element to build into the simulation, as businesses are of different sizes. This example will be specified using each of the approaches discussed above.

First, the number of employees for each of the eight businesses will be randomly generated using the `sample()` function. Suppose the sampling frame for the eight businesses ranges from 100 to 10,000 employees; this can be sampled using the `sample()` function as follows.

```
set.seed(38)

level1_ss <- sample(100:10000, size = 8, replace = TRUE)
```

This vector of eight business sizes can then be passed into the simulation arguments to generate unbalanced sample sizes across the eight businesses. A table of the sample sizes for each business is shown to highlight how these match the level one sample sizes generated above through the `sample()` function.

```
sim_arguments <- list(
  formula = culture ~ 1 + years_in_business + tech_company +
    (1 | business),
  fixed = list(
    years_in_business = list(var_type = "ordinal", levels = 2:20),
    tech_company = list(var_type = 'factor',
                        levels = c('Tech', 'Not Tech'))
  ),
  randomeffect = list(
    business_int = list(variance = 1.5, var_level = 2)
  ),
  error = list(variance = 10),
  reg_weights = c(60, -0.1, 2),
  sample_size = list(
    level1 = level1_ss,
    level2 = 8
  )
)
simulate_fixed(data = NULL, sim_arguments) |>
  dplyr::count(business)
```

Below are the other two ways that make it easy to specify sample sizes for each business. The first uses the `runif()` function with `round(0)` to ensure that the random numbers are whole integers. Almost any random generator could be used when rounding to whole integers, presuming that the sample sizes match the desired. The benefit of this would be to generate sample sizes more easily from different distributions (for example, skewed distribution of sample sizes). The second one below specifies the eight business sizes directly in a numeric vector, offering the most control over sample sizes.

```
set.seed(38)

level1_ss <- runif(8, min = 100, max = 10000) |>
  round(0)
```

```
level1_ss <- c(953, 5410, 101, 2432, 8760, 1012, 4431, 6755)
```

3.6.1.2 Unbalanced Level 3 Simulation

When simulating unbalanced data at levels one and two, the process is similar, except the sample sizes need to be specified for levels one and two. To do this, the sample sizes must be specified from the highest to the lowest. Therefore, the level three sample size needs to be identified first. Then, the level two sample sizes are specified for each level three unit. Lastly, the level one sample size is specified for each level two unit. For example, suppose the business culture example is extended to group businesses by the country they reside in. Suppose businesses can be sampled from four countries to send work culture surveys to. The specified level two sample size would reflect the number of businesses within each country to which surveys would be sent.

Any of the options discussed for specifying the level one sample sizes above can be used for the level two sample size here. In the example below, the object `level2_ss` is specified manually for each of the four countries. The sample sizes specified are the number of businesses within each country that are sent surveys. For example, three businesses are sent surveys in the first country, five in the second country, and so on.

```
level2_ss <- c(3, 5, 8, 4)
```

Now that the number of level three (countries) and level two (businesses within each country), the sample sizes for each business can be specified. The total number of businesses that the survey was sent to would be the sum of the number of businesses. In the example above, this would mean the sum of the `level2_ss` object, which should be 20 businesses. This represents the total number of level one (employees) sample sizes that need to be specified. Below, these are specified using the `runif()` function.

```
level1_ss <- runif(sum(level2_ss), min = 100, max = 10000) |> round(0)
```

These could then be combined to simulate the data. A table is generated that shows the number of employees for each business and country. The number of businesses within each country is also depicted. This confirms that the sample sizes specified accurately reflect what is being specified.

```
sim_arguments <- list(
  formula = culture ~ 1 + years_in_business + tech_company +
    (1 | business) + (1 | country),
  fixed = list(
    years_in_business = list(var_type = "ordinal", levels = 2:20),
    tech_company = list(var_type = 'factor',
                        levels = c('Tech', 'Not Tech'))
  ),
  randomeffect = list(
    business_int = list(variance = 1.5, var_level = 2),
    country_int = list(variance = 0.5, var_level = 3)
```

```
  ),
  error = list(variance = 10),
  reg_weights = c(60, -0.1, 2),
  sample_size = list(
    level1 = level1_ss,
    level2 = level2_ss,
    level3 = 4
  )
)
simulate_fixed(data = NULL, sim_arguments) |>
  dplyr::count(business, country) |>
  dplyr::group_by(country) |>
  dplyr::mutate(num_employee_country = sum(n),
                num_country = length(business))
```

3.7 Multiple Membership Nesting

Multiple membership nesting is a more complicated form of nesting in which level one units are neatly found within a single level two unit. Instead, some or all of the level one units below to more than a single level two unit. Let's frame this in an employee situation. As in the prior example, employees are nested within a single business if data are collected at a single measurement occasion. However, additional levels could be embedded within a business. For example, an employee could belong to multiple teams, such as a technology, human resources, researcher, or other type of team. An employee may belong to a single team, or they could belong to multiple teams. This would be an example of multiple membership and is a more complicated type of nesting.

When generating multiple member data with the **simglm** package, the critical additional piece of simulation arguments to pass is `multiple_member = TRUE` to the random effect argument for the associated term that identifies the multiple membership data. In the below example, this was specified for the term "secondaryteam." In addition, an optional argument was specified that reflects that ten secondary teams are available for employees to be a part of each company.

```
set.seed(12)

level2_ss <- c(30, 40, 50, 15, 12, 40, 75, 35)
level1_ss <- runif(sum(level2_ss), min = 100, max = 500)
```

```
sim_arguments <- list(
  formula = culture ~ 1 + years_in_business + tech_company +
    (1 | primaryteam) + (1 | secondaryteam) + (1 | business),
  fixed = list(
    years_in_business = list(var_type = "ordinal", levels = 2:20),
    tech_company = list(var_type = 'factor',
                        levels = c('Tech', 'Not Tech'))
  ),
  randomeffect = list(
    primaryteam_int = list(variance = 4, var_level = 2),
    secondaryteam_int = list(variance = 3, var_level = 2,
                             multiple_member = TRUE,
                             nids = 10),
    business_int = list(variance = 1.5, var_level = 3)
  ),
  error = list(variance = 10),
  reg_weights = c(60, -0.1, 2),
  sample_size = list(
    level1 = level1_ss,
    level2 = level2_ss,
    level2 = 8
  )
)
simulate_fixed(data = NULL, sim_arguments) |>
  dplyr::count(business)
```

3.8 Post-Processed Attributes

One type of attribute that is often included in models that have nesting is values that depend on an attribute at a lower level. For example, suppose data are collected where families are nested within city neighborhoods. In this case, family income was collected for everyone in a city. It could then be interesting to include or adjust for differences in neighborhood income. This attribute would be a function of the families' income within that specific neighborhood, an idea often referred to as contextual effects (Little et al., 2007; Raudenbush and Bryk, 2002). It should also be noted that special care is needed when interpreting these types of effects. As discussed in Gelman (2006), care needs to be taken when interpreting these types of effects as causal, and ecological fallacy should be considered (see also Wakefield (2003), cited within Gelman (2006)).

To generate this with simulated data, this attribute would not be random but a function of the random nature of the individual income within each neighborhood. This type of attribute is very different than one that would genuinely be random at the neighborhood level. For example, one could generate a neighborhood walkability score or neighborhood green space score that would not depend on family observations. This type of attribute would be generated just as other random attributes where the attribute level would be specified as the neighborhood level (or level two in this specific example).

In the case of the former, creating an attribute that is a function of another random attribute already simulated, the **single** package defines these are post-processed attributes. When generating a post-processed attribute, additional simulation arguments are specified. These are specified with the simulation argument name: "post." The syntax within the post-simulation arguments is slightly different from the data attributes that are truly random, specified within the "fixed" simulation arguments.

When specifying post-processing attributes, three required arguments are specified in a list for each post-processed attribute. The post-processed attributes are specified with a . First, the name of the attribute that was generated inside the "fixed" simulation arguments, named `variable`. This is the outcome or attribute of interest for which the resulting elements will be computed. Second, the name of the grouping attribute for which computations will be aggregated across is named `by`. Finally, a function that will be used to perform the computations is named `fun`. These types of attributes use the `aggregate()` function internally, which would generally look of the following form: `aggregate(variable, by, fun)`. The process for this type of attribute could be summarized as computing the function for the variable specified by unique combinations of the by attribute.

Using this logic with the post-processing simulation conditions below, the mean for the attribute "family_income" would be computed for each unique neighborhood. Therefore, the data will have 50 unique mean neighborhood income values. The code is not processed below, but the example is given to build off of.

```
sim_arguments <- list(
  formula = retirement_savings ~ 1 + age + family_income +
    walkability + neighborhood_income_post + (1 | neighborhood),
  fixed = list(
    age = list(var_type = 'ordinal', levels = 20:80),
    family_income = list(var_type = 'continuous', mean = 0, sd = 25000),
    walkability = list(var_type = 'continuous', mean = 35, sd = 20,
                       var_level = 2,
                       floor = 0, ceiling = 100)
  ),
  post = list(
```

```
    neighborhood_income_post = list(
      variable = 'family_income',
      by = 'neighborhood',
      fun = 'mean'
    )
  ),
  randomeffect = list(
    neighborhood_int = list(variance = 10000, var_level = 2)
  ),
  error = list(
    variance = 50000
  ),
  reg_weights = c(50000, 5000, 2, 1000, 0.5),
  sample_size = list(
    level1 = 1000,
    level2 = 50
  )
)
```

Before continuing the discussion, the mean is a common function for the computation of these types of attributes; however, the **simglm** allows the user to specify whichever function they would want. For example, the median could be helpful for a highly skewed attribute. User-written functions can also be specified. The only limitation to the function being used is that it must return a single value, not more than one value. For example, the `range()` function in R returns two values and is not a valid function, but the `diff()` function, which computes the difference between the minimum and maximum values, would be a valid function. Another post-processing functionality is built into the **single** package as well; this will be introduced in Chapter @ref{interrupt-ts}.

4

Simulating Generalized Linear (Mixed) Models

4.1 Chapter Overview

This chapter introduces simulation of generalized linear models and generalized linear mixed models for non-continuous outcomes. The focus is on binary outcomes modeled through logistic regression, though the framework generalizes to other outcome types. The chapter explains how link functions, linear predictors, and outcome distributions interact in generalized models and how simulation parameters correspond to meaningful quantities such as probabilities and odds ratios. Extensions to clustered and multilevel generalized models are presented, highlighting challenges unique to non-linear models, including interpretation and variance specification. Through applied examples, the chapter demonstrates how simulation can be used to evaluate power, bias, and model performance in settings where analytic solutions are limited. Emphasis is placed on helping applied researchers understand how assumptions in generalized models affect inference and study design decisions.

Moving beyond continuous outcomes and the general linear (mixed) model to generalized linear (mixed) models is possible with the **simglm** package. All of the code used in the previous two chapters can be used to simulate a generalized linear (mixed) model with **simglm**. The primary change results in one additional simulation argument named `outcome_type`.

4.2 Binary Outcomes

Binary or dichotomous outcomes can be simulated with the **simglm** package via a logistic regression model. Logistic regression models the probability of being in one of the two classes given a series of attributes. This model can be expressed as the following:

$$logit[Pr(y = 1)] = log(\frac{Pr(y = 1)}{1 - Pr(y = 1)}) = \alpha + \beta_k X \tag{4.1}$$

DOI: 10.1201/9781003177876-4

where the probability of the outcome equals one of the classes or groups, denoted as 1 in Equation (4.1), as a function of the linear predictors on the right-hand side of Equation (4.1). The linear predictor typically has an intercept, denoted as α in Equation (4.1), and k regression coefficients that represent the association of an attribute to the outcome on the logistic metric. Often, the regression coefficients are exponentiated to place the regression coefficients in the odds metric.

It is also often of interest, particularly for continuous attributes, to represent the probability change in the outcome for different values of the continuous attributes. This continuous probability can be computed as the following.

$$P(y = 1) = \frac{1}{1 + exp[-(\alpha + \beta_k X)]} \tag{4.2}$$

where the exp is the exponential function, and the other terms are defined with Equation (4.1).

Suppose we wanted to simulate a data situation where the outcome was curing a disease in a patient or not. This outcome would often be coded in a binary fashion where 1 = cured, 0 = otherwise. Suppose a pharmaceutical company is wishing to test how well a preliminary drug is effective in treating a disease. There may be doses being tested, and it is known that the patient's age is a known element that impacts efficacy. *Note*: This is a relatively simple example and may not (read, is likely not) be exactly how this process is done and is used solely for this specific example.

The following is a set of simulation arguments that would simulate an example similar to the one described above.

```
sim_arguments <- list(
  formula = cured ~ 1 + age + treatment,
  fixed = list(age = list(var_type = 'ordinal', levels = 20:80),
               treatment = list(var_type = 'factor',
                                levels = c('0 mg', '10 mg', '25 mg', '50 mg'))),
  sample_size = 1200,
  reg_weights = c(0.5, -.01, .1, .2, .25),
  outcome_type = 'binary'
)

simulate_fixed(data = NULL, sim_arguments) |>
  simulate_error(sim_arguments) |>
  generate_response(sim_arguments)
```

The simulated data contain the binary outcome, named "cured" here, and the untransformed outcome, named "untransformed_outcome". In the case of the binary outcome, the untransformed outcome is assumed to be on the logistic scale, and the **simglm** package uses `rbinom()` to convert these values to the binary 0/1 outcome. In **simglm**, `simulate_error()` adds noise to the linear predictor prior to applying the inverse-logit, which can be used to control extra variability or unexplained variance beyond the canonical logistic mean–variance relationship.

Recall that the regression weights specified with `reg_weights` correspond to the regression terms as specified in the formula argument. They are processed in the order produced by the `model.matrix()` function. In this case, these are the regression weights associated with each term:

- Intercept = 0.5
- Age = −0.01
- Treatment (10 mg) = 0.1
- Treatment (25 mg) = 0.2
- Treatment (50 mg) = 0.25

```
##   X.Intercept. age treatment_1 treatment_2 treatment_3
## 1            1  40           0           0           1
## 2            1  71           0           0           0
## 3            1  31           0           0           0
##   treatment level1_id   error fixed_outcome
## 1     50 mg         1  1.0084          0.35
## 2      0 mg         2 -0.0984         -0.21
## 3      0 mg         3 -0.0400          0.19
##   random_effects untransformed_outcome cured
## 1              0                1.3584     1
## 2              0               -0.3084     1
## 3              0                0.1500     1
```

4.2.1 Adding a Random Effect

Adding a random effect would follow the same structure as in Chapter 3. Simulation arguments would include the random effects specified in the formula and as a named "randomeffect" element of the simulation arguments. Finally, an additional level of sample size would be specified. In the following example, there may be a hospital or regional effect that could be added. More specifically, this could reflect a situation where hospitals were chosen randomly to be part of the clinical trial and would represent a random portion of variation, and the goal would be for the treatment to be effective outside of the current sample of hospitals involved in the clinical trial.

One note, the random effect variance was specified as equal to 3. This would be large for this logistic regression and we can estimate how much variance is attributable to hospitals by approximating the intra-class correlation (ICC). An approximation can be:

$$ICC \approx \frac{\sigma_u^2}{\sigma_u^2 + \pi^2/3} \tag{4.3}$$

where σ_u^2 is the random effect variance and $\pi^2/3$ is the logistic latent-scale variance. In this example, $ICC \approx \frac{3}{3+3.48} \approx 0.48$, assuming $\pi^2/3 \approx 3.48$, suggesting a significant amount of between-hospital heterogeneity. This is intentionally large for illustration; in many applications it would imply very strong between-hospital heterogeneity.

```
sim_arguments <- list(
  formula = cured ~ 1 + age + treatment + (1 | hospital),
  fixed = list(age = list(var_type = 'ordinal', levels = 20:80),
               treatment = list(var_type = 'factor',
                                levels = c('0 mg', '10 mg',
                                           '25 mg', '50 mg'))),
  randomeffect = list(
    int_hospital = list(variance = 3, var_level = 2)
  ),
  sample_size = list(level1 = 100, level2 = 20),
  reg_weights = c(0.5, -.01, .1, .2, .25),
  outcome_type = 'binary'
)

simulate_fixed(data = NULL, sim_arguments) |>
  simulate_error(sim_arguments) |>
  simulate_randomeffect(sim_arguments) |>
  generate_response(sim_arguments)
```

The simulated data for this example are not explicitly shown but are left for the reader to run independently. The difference between this example with the random effect and the prior example is the addition of the random effect column and the multilevel nature of the data.

4.3 Ordinal Outcomes

Ordinal outcomes can also be simulated, both with and without random effects. Ordinal outcomes are a generalization of logistic regression to more than two categories. In the ordinal models discussed here, it is assumed that the outcome can be ordered in some fashion rather than simply being categorical attributes that do not offer any order. For example, rating scale items from strongly disagree, disagree, agree or strongly agree would be considered an ordinal attribute. With this type of data, it is often assumed that a latent or unobserved attribute is being approximated or measured through regions separated by thresholds (Fox, 2015).

The logistic distribution transforms the ordinal models considered in the **simglm** package. The statistics literature also uses the probit distribution for this model type; however, the **simglm** only supports the logistic distribution. Following the parameterization found in Fox (2015) (see equation 14.28), an ordinal simulation model could look as follows:

$$logit[Pr(y > j)] = \alpha_j + \beta_k X, for\ j = 1,\ 2,\ ...,\ m - 1 \tag{4.4}$$

where j refers to the thresholds that contrast the categories above j to those below category j. The subscript k refers to the number of attributes specified in the design matrix, X (i.e., the number of columns in X). Equation (4.4) is sometimes referred to as the proportional odds model due to the data condition that the slope terms are the same across the cumulative logits. The β_k are assumed to be equal across the different thresholds. This can be relaxed by estimating some or all of the slopes across the cumulative logit comparisons. This more flexible model could be written as:

$$logit[Pr(y > j)] = \alpha_j + \beta_{jk} X, for\ j = 1,\ 2,\ ...,\ m - 1 \tag{4.5}$$

For this type of model, both the slopes and the intercepts vary across different thresholds.

When specifying ordinal outcomes in **simglm**, the simulation argument, `outcome_type = 'ordinal'`, is specified. The `reg_weights` simulation argument must be specified as a list when this is specified. Within the list, a vector of regression coefficients is provided for each cumulative logit. In this scenario, the number of categories for the ordinal outcome would be one more than the number of vectors specified in the `reg_weights` argument.

The code below specifies four vectors within the list of `reg_weights` simulation arguments. Therefore, five ordinal categories will be simulated. In the `reg_weights` argument, you may also notice that only the first number varies across the vectors. The first number represents the α_j term from Equation (4.4) of the proportional odds model. Different software uses different signs/definitions for cutpoints; **simglm** follows the specification by Fox (2015), so the values supplied here correspond to negative thresholds under that convention. Therefore, the intercepts get smaller from the first vector to the last. Suppose the model was fitted (e.g., with `ordinal::clm()`), the thresholds would be the specified negative intercept coefficient. This model also assumes the proportional odds across the cumulative logits due to the same slope coefficients being specified.

```
set.seed(2022)

sim_arguments <- list(
```

```
  formula = y ~ 1 + turnover + treatment,
  fixed = list(turnover = list(var_type = 'continuous',
                               mean = 10, sd = 3),
               treatment = list(var_type = 'factor',
                                levels = c('control',
                                           'treatment'))
  ),
  error = list(variance = 0.1),
  sample_size = 1000,
  reg_weights = list(
    c(0.5, 0.2, -0.35),
    c(-1, 0.2, -0.35),
    c(-2, 0.2, -0.35),
    c(-5, 0.2, -0.35)
  ),
  outcome_type = 'ordinal'
)

simulate_fixed(data = NULL, sim_arguments) |>
  simulate_error(sim_arguments) |>
  generate_response(sim_arguments)
```

4.3.1 Relaxing Proportional Odds Condition

The proportional odds condition can be relaxed using Equation (4.5), which allows one or more of the slope coefficients to vary in addition to the intercepts that already vary. Within the **simglm** package, this can be achieved by allowing one of the slope terms to vary instead of being fixed. The following code is an example of the generating model being one in which the proportional odds condition is being specified for the treatment attribute (slope fixed at -0.35). However, the attribute turnover has differential effects across the cumulative logits, varying from 0.25, 0.1, -0.1, and 0.

```
set.seed(2022)

sim_arguments <- list(
  formula = y ~ 1 + turnover + treatment,
  fixed = list(turnover = list(var_type = 'continuous',
                               mean = 10, sd = 3),
               treatment = list(var_type = 'factor',
                                levels = c('control',
                                           'treatment'))
  ),
```

```
  error = list(variance = 0.1),
  sample_size = 1000,
  reg_weights = list(
    c(0.5, 0.25, -0.35),
    c(-1, 0.1, -0.35),
    c(-2, -0.1, -0.35),
    c(-5, 0, -0.35)
  ),
  outcome_type = 'ordinal'
)

simulate_fixed(data = NULL, sim_arguments) |>
  simulate_error(sim_arguments) |>
  generate_response(sim_arguments)
```

4.3.2 Add in a Random Effect

Adding a random effect is similar to that of other models. The code below adds a random intercept to the proportional odds logit model in which individuals would be nested within hospitals.

```
set.seed(2022)

sim_arguments <- list(
  formula = y ~ 1 + turnover + treatment + (1 | hospital),
  fixed = list(turnover = list(var_type = 'continuous',
                               mean = 10, sd = 3),
               treatment = list(var_type = 'factor',
                                levels = c('control',
                                           'treatment'))
  ),
  randomeffect = list(int_hospital = list(
    variance = 3, var_level = 2)
  ),
  error = list(variance = 0.1),
  sample_size = list(level1 = 30, level2 = 20),
  reg_weights = list(
    c(0.5, 0.2, -0.35),
    c(-1, 0.2, -0.35),
    c(-2, 0.2, -0.35),
    c(-5, 0.2, -0.35)
  ),
  outcome_type = 'ordinal'
```

```
)

simulate_fixed(data = NULL, sim_arguments) |>
  simulate_error(sim_arguments) |>
  simulate_randomeffect(sim_arguments) |>
  generate_response(sim_arguments)
```

4.4 Multinomial Outcomes

Multinomial outcomes are those with more than two categories, but those categories are not ordinal. Instead, these categories are unordered and represent different groups or categories. In this situation, this type of model is often referred to as multinomial regression (Fox, 2015).

Multinomial regression is defined by first setting a reference category for which the likelihood of being in another category is compared. This type of model can be defined as:

$$logit[\frac{Pr(y = c)}{Pr(y = r)}] = \beta_{ck} X, for\ c\ 1,\ 2,\ ...,\ m - 1 \tag{4.6}$$

where c is a specific category for the outcome, and r is the reference category for the outcome. The c subscript indexes from 1 to $m - 1$ where m is the total number of groups in the outcome. This means that the number of comparisons to the reference group is one less than the total number of groups in the outcome attribute. For example, if the outcome attribute has three groups (e.g., Depression, Anxiety, or ADHD diagnoses), then two would be compared to the reference category.

Finally, the β_{ck} are the regression coefficients that estimate the relationship between the attributes included in the X matrix and the outcome comparisons. These regression coefficients are typically defined in the logit metric; therefore, exponentiating these coefficients would turn them into the odds metric. Finally, the c subscript in β_{ck} represents which category comparison is being made, and k indexes the number of attributes or columns in the design matrix, X. Notice that the regression coefficient vector in this model, β_{ck}, indexes by the category comparison being made. This is similar to the case for the ordinal outcome, where the proportional odds assumption was relaxed to estimate unique regression coefficients for each comparison. This is the default behavior in the multinomial regression, where unique regression coefficient estimates are obtained for each category comparison made. In the example above, with three diagnoses, two unique sets of regression coefficients would be estimated, given that there are $m - 1$ comparisons made.

4.4.1 Multinomial Simulation with simglm

Within the **simglm** package, multinomial regression simulation is similar to the ordinal simulation. The primary difference is that the outcome type simulation argument is specified as `outcome_type = 'multinomial'`, instead of the `outcome_type = 'ordinal'`. Similar to the ordinal simulation, a list of population regression coefficients represents the relationship between the attributes in the model equation and the outcome. The list length should be one less than the number of categories in the desired outcome. For example, in the three diagnosis examples above, the list of regression coefficients would contain two vectors of regression coefficients. Finally, an optional but recommended argument called `multinomial_categories` is specified to give the outcome category names instead of being represented by numbers. A complete example below is shown for the three category mental health diagnosis example used.

```
set.seed(2023)

sim_arguments <- list(
  formula = diagnosis ~ 1 + age + season,
  fixed = list(age = list(var_type = 'continuous', mean = 20, sd = 2),
               season = list(var_type = 'factor',
                             levels = c('winter', 'summer'))),
  error = list(variance = 0.5),
  sample_size = 1000,
  reg_weights = list(
    c(-3, 0.2, -1),
    c(-1, 0.1, -0.8)
  ),
  outcome_type = 'multinomial',
  multinomial_categories = c('ADHD', 'Anxiety', 'Depression')
)

tmp_data <- simulate_fixed(data = NULL, sim_arguments) |>
  simulate_error(sim_arguments) |>
  generate_response(sim_arguments)
```

Notice in this example that two attributes are being generated to be related to the diagnosis. These include the individual's age and the season of the evaluation for diagnosis (a factor specified as either summer or winter). Secondly, also notice that there are two vectors of regression coefficients within the `reg_weights` simulation argument, which is one less than the number of categories specified within the `multinomial_categories` simulation argument.

Notice that in this example, different values are specified for each vector of the regression coefficients. This is the default estimation for the multinomial model. However, if there was a theory suggesting the parameters should be the same across the comparisons, these could be specified as equal.

In this example, Depression is treated as the reference category. The first element of `reg_weights` represents the log-odds of ADHD versus Depression, and the second represents the log-odds of Anxiety versus Depression.

4.4.1.1 Adding Random Effect

It is also possible to add random effects to the multinomial simulation example. This follows a similar discussion of random effects. To add random effects, the **lme4** R package type syntax is used to specify those. In the example below, the clinic level effect was added to suggest that patients within a clinic would be more similar than those across different clinics. Finally, when adding this type of random effect, the **simglm** package assumes the data should be generated that are nested, where individuals would belong to specific clinics. Therefore, the sample size needs to specify the number of individuals within a clinic (i.e., `level1 = 15`) and the number of clinics (i.e., `level2 = 20`). As such, this simulation would be balanced with 15 individuals evaluated at 20 different clinics. The random effect would have a variance of 3 in the logit metric.

4.5 Count Outcomes

Similar to binary outcomes, count outcomes can be simulated using the `outcome_type = 'count'` or `outcome_type = 'poisson'` simulation argument. When the outcome type is specified as count or poisson, the `rpois()` function transforms to the number outcome. These make the assumption that the variance equals the mean from the Poisson distribution. It is possible to add additional unexplained variance to the untransformed outcome via the `simulate_error()` specification. Adding error to the linear predictor induces overdispersion (a Poisson–lognormal-type mixture) and if it is not specified in the simulation arguments, the default behavior in **simglm** makes the error variance equal to one.

5

Varying Simulation Parameters

5.1 Chapter Overview

This chapter introduces systematic approaches to varying simulation parameters in order to explore design sensitivity and robustness. Rather than focusing on a single set of assumptions, the chapter demonstrates how grids of parameter values can be used to examine how results change across plausible scenarios. The chapter discusses practical strategies for varying effect sizes, sample sizes, variance components, and model structures, as well as how to organize and summarize results across simulation conditions. Examples illustrate how parameter variation can reveal non-linearities, threshold effects, and tradeoffs that are not apparent from single-scenario analyses. The chapter emphasizes the importance of exploring uncertainty in design assumptions and positions parameter variation as a core advantage of simulation-based approaches.

It is possible to generate multiple simulated data with a single set of simulation arguments using the **simglm** package. There can be many uses for this, such as:

- Conduct a power analysis
- Evaluate statistical properties of an estimator/method
- Generate multiple copies for a class with different properties

Varying parameters when simulated data for a power analysis or to evaluate statistical properties is done to understand the impact different conditions have on the simulated data. In these situations, a model is typically also fitted to the data; this will be shown when conducting a power simulation in Chapters 8 and 9. This chapter will focus on simply varying simulation parameters for generating data with different properties.

5.2 Varying Parameters Syntax

The element to vary simulation conditions is through the named simulation argument, `vary_arguments`. Different simulation arguments can be specified

DOI: 10.1201/9781003177876-5

inside this named simulation argument containing multiple different simulation arguments. To be more concrete, a common scenario to vary in simulated data is the sample size. Suppose we wanted to simulate data with 50 rows of data and another with 150 rows of data.

```
sim_arguments <- list(
  formula = amount ~ 1 + treat,
  fixed = list(
    treat = list(var_type = 'factor',
                 levels = c('Control', 'Treatment'))
  ),
  reg_weights = c(0, 0.5),
  error = list(variance = 2),
  vary_arguments = list(
    sample_size = list(50, 150)
  )
)
```

Notice how there is no direct specification of the `sample_size` argument within the simulation arguments. Instead, this simulation argument is specified within the `vary_arguments` simulation argument. The sample size is specified similarly to how it was specified before.

Using the `vary_arguments` syntax would be equivalent to specifying the simulation arguments twice. For example, this would generate similar simulated data, but one would generate data with 50 data units and the second would have 150.

```
sim_arguments <- list(
  formula = amount ~ 1 + treat,
  fixed = list(
    treat = list(var_type = 'factor',
                 levels = c('Control', 'Treatment'))
  ),
  reg_weights = c(0, 0.5),
  error = list(variance = 2),
  sample_size = 50
)

sim_arguments<- list(
  formula = amount ~ 1 + treat,
  fixed = list(
    treat = list(var_type = 'factor',
                 levels = c('Control', 'Treatment'))
```

```
  ),
  reg_weights = c(0, 0.5),
  error = list(variance = 2),
  sample_size = 150
)
```

The `vary_arguments` specification does this expansion for you inside of the `replicate_simulation()` function, which can lead to more compact syntax with less duplication. This is often a hallmark of functional programming and can help to reduce errors, make debugging easier, and make it easier to read the resulting code.

To generate data with the `vary_arguments` specified, a different function, `replicate_simulation()` is used. By default, a single replication is performed. Think of a replication as the number of times each set of simulation arguments creates a simulated data set. If the number of replications is set to more than one, then the same set of simulation conditions would generate more than one data set.

```
sim_arguments <- list(
  formula = amount ~ 1 + treat,
  fixed = list(
    treat = list(var_type = 'factor',
                 levels = c('Control', 'Treatment'))
  ),
  reg_weights = c(0, 0.5),
  error = list(variance = 2),
  vary_arguments = list(
    sample_size = list(50, 150)
  )
)

replicate_simulation(sim_arguments)
```

5.2.1 Varying More Than One Simulation Condition

It is possible to vary more than a single simulation condition using the `vary_arguments` simulation argument. This structure is similar to specifying the other simulation arguments; these are named elements embedded within the `vary_arguments` list. The following is a general structure of what is expected when more than one simulation argument is specified.

```
sim_arguments <- list(
  ...,
  vary_arguments = list(
    sample_size = list(...),
    error = list(...),
    reg_weights = list(...)
  ),
  ...
)
```

In this structure, each simulation argument that is desired to be varied is included by name within `vary_arguments`. Each named element is a list. Within each list, the same structure as the original simulation argument is described in earlier chapters of the book. For example, to specify the error simulation argument, a nested list is created where each element of the nested list is one error specification. Below is an example of what this would look like.

```
sim_arguments <- list(
  ...,
  vary_arguments = list(
    sample_size = list(...),
    error = list(
      list(variance = 1),
      list(variance = 10)
    ),
    reg_weights = list(...)
  ),
  ...
)
```

Adding the additional simulation arguments are specified similarly. The regression weights with `reg_weights` would be vectors of the same length that would match the number of regression parameters that were included in the model formula. Secondly, the sample size is included as shown above, where each element within the list is a different sample size for the regression model.

```
sim_arguments <- list(
  ...,
  vary_arguments = list(
    sample_size = list(50, 150),
    error = list(
      list(variance = 1),
      list(variance = 10)
```

```
    ),
    reg_weights = list(
      c(0, 0.5),
      c(1, 0.5),
      c(0, 1),
      c(1, 1)
    )
  ),
  ...
)
```

Notice in the vary arguments above, any number can be included within each one. These are combined in a factorial manner using the `expand.grid()` function in R. In the case above, there are two sample size conditions, two error conditions, and four regression weight conditions. Within a factorial expansion, this would result in $2 * 2 * 4 = 16$ total simulation conditions. Table 5.1 shows the crossed simulation conditions that would be simulated using the `replicate_simulation()` function.

TABLE 5.1 The factorial expansion of the varied simulation conditions.

sample_size	error	reg_weights
50	1	0.0, 0.5
150	1	0.0, 0.5
50	10	0.0, 0.5
150	10	0.0, 0.5
50	1	1.0, 0.5
150	1	1.0, 0.5
50	10	1.0, 0.5
150	10	1.0, 0.5
50	1	0, 1
150	1	0, 1
50	10	0, 1
150	10	0, 1
50	1	1, 1
150	1	1, 1
50	10	1, 1
150	10	1, 1

5.3 Replicating the Simulated Data

The setting of replications is another named simulation argument. In the prior example, being more explicitly in the simulation arguments would be adding the number of replications explicitly. For example, the following is a more explicit (and transparent) version of the previous section. The replications state that one data set should be simulated for a sample size of 50 and 150.

```
sim_arguments <- list(
  formula = amount ~ 1 + treat,
  fixed = list(
    treat = list(var_type = 'factor',
                 levels = c('Control', 'Treatment'))
  ),
  reg_weights = c(0, 0.5),
  error = list(variance = 2),
  vary_arguments = list(
    sample_size = list(50, 150)
  ),
  replications = 1
)

replicate_simulation(sim_arguments)
```

If more data sets within each set of simulation conditions should be generated, then the replications simulation argument can be specified to a numeric value larger than one. For example, if 10 data sets for each set of simulation conditions were desired, the replications simulation argument could be set to 10.

In practice, when estimating power or evaluating the statistical properties of a model or estimator, many more replications would be desirable. The number of replications is similar to the sample size of an empirical study, where greater precision in estimating the outcome is gained by increasing the number of replications. If some precision is desired, then at least 1,000 replications would likely be needed, but increasing to 5,000 or 10,000 would be even better. Monte Carlo standard errors are discussed in more detail in Chapter 10.

5.3.1 Replication Without Varying Arguments

It is unnecessary to vary arguments to generate more than one data set from a single set of simulation arguments. Suppose you were teaching a class or workshop and wanted each student to have slightly different data but with the same sample size and being simulated from the same set of population values. One set of simulation arguments could be defined, and then the `replicate_simulation()` function could be used with the number of replications specified.

Suppose there were 30 students in the class or workshop; then the following could be used to generate 30 unique data sets with the same underlying population.

```
sim_arguments <- list(
  formula = amount ~ 1 + treat,
  fixed = list(
    treat = list(var_type = 'factor',
                 levels = c('Control', 'Treatment'))
  ),
  reg_weights = c(0, 0.5),
  error = list(variance = 2),
  sample_size = 100,
  replications = 30
)

replicate_simulation(sim_arguments)
```

6

Simulating Data That Are Realistic

6.1 Chapter Overview

This chapter focuses on designing simulations that reflect realistic data conditions encountered in applied research. Topics include incorporating missing data, non-normal error distributions, skewed covariates, and measurement error into the data-generating process. The chapter discusses different missingness mechanisms and their implications for simulation-based power analysis and model evaluation. Practical guidance is provided on grounding simulation parameters in prior research, pilot data, or domain knowledge. By emphasizing realism, the chapter shows how simulation can be used not only to assess idealized designs but also to anticipate challenges that arise in practice. The goal is to help researchers design studies and analyses that are robust to departures from simplifying assumptions.

Simulating data that reflect realistic conditions is especially useful when simulation results are intended to inform decisions about studies conducted in the real world. One common use case is simulation-based power analysis, where the goal is to generate data that resemble what will eventually be collected. Because those data do not yet exist, the data-generating process must be grounded in what is already known, prior research, domain expertise, and, when available, pilot or historical data.

These sources can inform key elements of the simulation design, particularly the magnitude of effects. Standardizing predictors is often helpful because regression weights can then be interpreted as standardized effects. Standardization can also simplify the specification of residual variability: when predictors have variance one, the error variance can often be chosen more directly.

Even with strong prior evidence, the population effect size is rarely known with certainty. For this reason, it is often preferable to define a grid of plausible effect sizes (and other parameters) and examine how results change across that grid. In the context of power analysis, this approach naturally connects to the idea of power curves from traditional analytic power calculations.

In addition to effect magnitude, realistic simulations often incorporate other features commonly observed in real data, such as heterogeneity in outcome

DOI: 10.1201/9781003177876-6

variability, missingness, and correlation among predictors and/or random effects. Each of these considerations is discussed in the sections that follow.

6.2 Specifying Magnitude of Effect

This is the most important consideration when trying to simulate realistic data. There can be a few sources to consult when identifying the magnitude of the effect to be included in the simulation design. These sources could include some or all, of the following:

1. Recent meta-analyses (ideally combining causal research designs)
2. Recent empirical research utilizing a causal design
3. Recent empirical research utilizing non-causal designs

Each of these sources will be discussed in more detail. It should be noted, however, that generally, the sources are ordered from those that offer the most information to the least information to help identify the population effect of interest.

6.2.1 Utilizing Meta-Analyses

Meta-analyses quantitatively synthesize effect sizes from primary empirical studies for a topic of interest. If implemented well, meta-analyses can provide strong evidence for a likely range of true effect sizes by combining statistical information from multiple empirical studies. This can be compelling information when the empirical studies combined implemented causal study designs (e.g., studies that employ randomization), but it can still be impactful when combining correlational study designs (e.g., studies without randomization).

One reason meta-analyses can be a great way to identify the true effect size is that they can provide strong evidence for the heterogeneity of the effect size. Put another way, meta-analyses estimate how much the effect sizes vary across different empirical studies and weigh the average effect size based on the size of the empirical study. This weighted average effect size is a statistical means to give more weight to studies with increased precision levels. By combining multiple studies of varying sizes, meta-analyses can offer a more holistic view of the research literature to hone in on the population effect size.

Using a meta-analysis's weighted average effect size can be a great way to identify a starting place when simulating data. However, similar to the primary studies that are synthesized in a meta-analysis, this is only an estimate of the population effect size. Therefore, it is important to vary the effect size to incorporate the sampling error in the effect size estimate when simulating realistic data conditions. This idea will be discussed in more detail later in this book when using simulation to estimate statistical power. Incorporating information about the variability in the weighted average effect size by simulating population effect sizes of different magnitudes would be important to explore.

6.2.2 Utilizing Causal Research Designs

A meta-analysis may have yet to be completed for new or emerging research literature on a research topic. Alternatively, a meta-analysis may have synthesized non-causal research designs. In these cases, a recent empirical study that utilized a causal research design, such as a randomized control trial, difference in difference, interrupted time series, or other strong research designs, may best estimate the effect size. Even if a meta-analysis was completed, it would likely be of interest to utilize a new causal primary research study to explore if the estimated effect size has shifted since the completion of the meta-analysis.

Causal research designs offer strong research control to isolate the desired effect size. As such, these studies can offer strong evidence of the magnitude of the effect size. These studies also estimate the sampling error to show how much the effect size may shift to a new sample from the same population. As such, all things being equal, a larger causal study will estimate the population effect size more precisely. Similar to when utilizing a meta-analysis, it would be important to consider the sampling error when simulating realistic data, particularly in estimating power for the effect size. It should also be noted that causal designs are excellent for estimating the population effect size for the population being studied. Suppose there is a mismatch between the population from which the sample was drawn in the causal study and the population being simulated. In that case, caution needs to be taken to use the estimated effect size for the new population. Causal designs have strong internal validity but do not necessarily have strong external validity (i.e., generalization).

6.2.3 Utilizing Non-causal Research Designs

Non-causal research designs should be used cautiously when estimating the population effect size. These designs are not intended to accurately estimate the population effect size (i.e., without bias). As such, these designs should only be used as a last resort when finding a value for the population effect size. If non-causal designs are the only possible or available for a given research topic, then varying effect sizes would be important to explore. This would be particularly important when estimating power.

6.3 Adding Conditional Variance and Treatment Effect Heterogeneity

In many real-world studies, variability in outcomes differs systematically across groups or across values of key attributes. One common example arises in treatment settings, where treated units often exhibit greater variability in outcomes than control or business-as-usual groups (Kent et al., 2010). This increased variability may reflect heterogeneous responses to the treatment, differences in implementation, or varying levels of engagement or compliance.

This type of variability differs from classical heteroskedastic error variance, which refers specifically to differences in the variance of the unobserved error term conditional on predictors. Instead, the variability considered here is conditional outcome variance. That is, outcome dispersion that depends on observed attributes such as treatment assignment. This phenomenon is closely related to treatment effect heterogeneity, as variability in outcomes often reflects differential responses to treatment across units. It can have important implications for estimation, inference, and statistical power.

Incorporating conditional variance heterogeneity into simulations can be especially informative for power analysis. Even when average treatment effects are unchanged, increased variability in certain groups can inflate standard errors and reduce power, particularly for models that implicitly assume homogeneous variance across groups.

Using **simglm**, conditional variance heterogeneity can be generated using the `simulate_heterogeneity()` function paired with the "heterogeneity" simulation argument. This approach allows the variability of the outcome to depend on the values of a specified attribute, such as treatment group membership or a continuous covariate.

The example shown below simulated heterogeneity about the "group" attribute. When specifying heterogeneity for a factor type fixed attribute, the variance for each group is specified. Care needs to be used when specifying the variance for each level, as the levels of the factor attribute are specified in alphabetic order for the variance heterogeneity arguments. To be more explicit, in this case, the variance equal to one would be for the "No Sunscreen" group as that factor level is closer to "A" in alphabetic order. Subsequently, the "Sunscreen" group would have a variance of eight.

One note about the process for the `simulate_heterogeneity()` function: this function needs to be specified after generating the fixed effects with the `simulate_fixed()` function and after simulating the random error with the `simulate_error()` function.

```
simulation_arguments <- list(
  formula = y ~ 1 + group,
  fixed = list(group = list(var_type = 'factor',
                            levels = c('No Sunscreen', 'Sunscreen'))),
  sample_size = 500,
  error = list(variance = 1),
  heterogeneity = list(variable = 'group',
                       variance = c(1, 8)),
  reg_weights = c(0, .15)
)

simulate_fixed(data = NULL, simulation_arguments) |>
  simulate_error(simulation_arguments) |>
  simulate_heterogeneity(simulation_arguments) |>
  head()
```

```
##   X.Intercept. group_1        group level1_id    error
## 1            1       1    Sunscreen         1 -5.14779
## 2            1       1    Sunscreen         2  2.53277
## 3            1       0 No Sunscreen         3  0.27271
## 4            1       0 No Sunscreen         4 -0.01891
## 5            1       1    Sunscreen         5 -1.09718
## 6            1       1    Sunscreen         6 -3.03023
##   orig_error
## 1   -1.82002
## 2    0.89547
## 3    0.27271
## 4   -0.01891
## 5   -0.38791
## 6   -1.07135
```

As can be seen from the output, the `simulate_heterogeneity()` function creates a new column representing the original error named "orig_error." This error is generated from the `simulate_error()` function and renamed. The `simulate_heterogeneity()` creates the column "error" representing the new error with the variance heterogeneity added to this error term. It is important to note that this approach induces systematic differences in outcome variability rather than purely random heteroskedastic errors. As such, it is best interpreted as simulating conditional variance or treatment effect heterogeneity rather than classical heteroskedasticity in the error term.

6.3.1 Conditional Variance for Continuous Attributes

When simulating variance heterogeneity for a continuous attribute, the `simulate_heterogeneity()` function splits the continuous attribute into the same number of groups based on the number of variance values specified. When looking at the `heterogeneity` simulation argument below, there are a total of five variance values specified; therefore, the "age" attribute will be split into five categories, and the variance values are specified for those ranges in order from smallest to largest. For the current example, this would mean that younger ages would have smaller variance values, increasing as age increases.

```
simulation_arguments <- list(
  formula = y ~ 1 + age,
  fixed = list(age = list(var_type = 'ordinal',
                          levels = 20:80)),
  sample_size = 500,
  error = list(variance = 1),
  heterogeneity = list(variable = 'age',
                       variance = c(1, 2, 4, 8, 16)),
  reg_weights = c(0, .15)
)

simulate_fixed(data = NULL, simulation_arguments) |>
  simulate_error(simulation_arguments) |>
  simulate_heterogeneity(simulation_arguments) |>
  head()
```

```
##   X.Intercept. age level1_id   error orig_error
## 1            1  28         1 -0.1687    -0.1687
## 2            1  25         2 -1.1976    -1.1976
## 3            1  71         3  3.9673     0.9918
## 4            1  21         4  0.2960     0.2960
## 5            1  42         5 -1.6914    -1.1960
## 6            1  20         6  0.7409     0.7409
```

6.4 Adding Missing Data

Missing data in real world data is almost always present and can impact the analysis (Little and Rubin, 2019; Rubin, 1976). The impact missing data can have depends on how the data are missing and how much missing data is (Schafer and Graham, 2002). The impact of missing data can range from decreasing precision to biased estimates entirely (Little, 1988). As such, it is important, particularly when estimating power, to consider the impacts that missing data can have on the resulting analysis (Graham, 2009). Simulating missingness lets you study the consequences; whether bias occurs depends on both the missingness mechanism and the analysis strategy.

Fortunately, the **simglm** package offers a variety of missing data simulation features. The package allows for the simulation of the following types of missing data (Diggle, 2002; Little and Rubin, 2019; Schafer and Graham, 2002):

- Random Missing Data, sometimes referred to as Missing Completely at Random (MCAR)
- Missing at Random (MAR)
- Drop out missing data, which is most useful for longitudinal designs.

Examples of these types of missing data generation will be discussed in more detail.

Of note, it is beyond the scope of this book to discuss types of missing data mechanisms; instead, interested readers are directed to several useful resources on that topic (Enders, 2022; Little and Rubin, 2019; Schafer and Graham, 2002).

6.4.1 Random Missing Data

Random missing data, or missing completely at random (MCAR), is a missing data mechanism where the missing is random. That is, no other factors are leading to the missing data. For example, the missing data is independent of other known factors (e.g., income, socio-economic status, etc.) or unknown factors. This is often the simplest type of missing data in that it can be simulated using random numbers to identify which data should be missing.

The following example simulates a linear regression model where the outcome is a function of an intercept, age, and treatment effect. The missing data arguments are specified within the `missing_data` simulation argument. The random missing data is specified inside this with the argument `type = 'random'`. The other required arguments for random missing data include `miss_prop` and `new_outcome`. The `miss_prop` argument is a numeric value that specifies the proportion of missing data that should be included in the new outcome that will have missing data. The `new_outcome` argument is the name of the column of data that will have the missing data included. The original simulated outcome without missing data will be retained in the simulated data for reference.

The function `generate_missing()` generates the missing data. This function must be called **after** the `generate_response()` function and other relevant `simulate_*()` functions. This is because the **simglm** package first simulates the complete data with no missing values, then generates a new column that has the missing data based on the missing data simulation arguments.

```r
set.seed(321)

sim_arguments <- list(
  formula = y ~ 1 + age + treat,
  reg_weights = c(0.75, 0, 0.33),
  fixed = list(age = list(var_type = 'ordinal', levels = 30:60),
               treat = list(var_type = 'factor',
                            levels = c('Treatment', 'Control'))
               ),
  missing_data = list(miss_prop = .25, new_outcome = 'y_missing',
                      type = 'random'),
  sample_size = 150
)

simulate_fixed(data = NULL, sim_arguments) |>
  simulate_error(sim_arguments) |>
  generate_response(sim_arguments) |>
  generate_missing(sim_arguments) |> head()
```

```
##   X.Intercept. age treat_1     treat level1_id
## 1            1  51       0 Treatment         1
## 2            1  47       0 Treatment         2
## 3            1  58       0 Treatment         3
## 4            1  42       1   Control         4
## 5            1  54       1   Control         5
## 6            1  53       1   Control         6
##      error fixed_outcome random_effects       y
## 1  0.76818          0.75              0  1.5182
## 2  0.01791          0.75              0  0.7679
## 3  1.77459          0.75              0  2.5246
## 4  0.67962          1.08              0  1.7596
## 5 -1.88092          1.08              0 -0.8009
## 6 -0.85548          1.08              0  0.2245
##   miss_prob miss_prob y_missing
## 1     0.949         0    1.5182
## 2     0.984         0    0.7679
## 3     0.093         1        NA
## 4     0.317         0    1.7596
## 5     0.783         0   -0.8009
## 6     0.243         1        NA
```

6.4.2 Missing at Random (MAR)

Missing data that are considered missing at random depend on one or more known attributes in the data.

The missing data mechanisms are specified similarly using the `missing_data` arguments. Within these simulation arguments, the name of the missing data outcome is specified in "new_outcome" as a character string. The missing data type is called "mar", standing for missing at random. When the type is "mar," then two additional missing data arguments are needed, `miss_cov` and `mar_prop`. The `miss_cov` specifies which attribute the missing data depends on. The current specification allows one attribute to be specified from the model formula. Once the missing attribute is specified, the `mar_prop` argument specifies the proportion of missing data based on different attribute values. A larger proportion would indicate that more missing data would occur.

How the proportion of missing data is specified depends on the type of attribute depicted in the `miss_cov` argument. If the attribute represents categories, then the same number of proportions as categories need to be specified. The **simglm** package preserves factor order when simulating categorical or factor attributes. Therefore, the order specified within the fixed argument would be the same order that the missing proportions in `mar_prop` would follow.

```
set.seed(321)

sim_arguments <- list(
  formula = y ~ 1 + age + treat,
  reg_weights = c(0.75, 0, 0.33, 0.6),
  fixed = list(age = list(var_type = 'ordinal', levels = 30:60),
               treat = list(var_type = 'factor',
                            levels = c('Control', 'Low Dose',
                                       'High Dose'))),
  missing_data = list(new_outcome = 'y_missing', miss_cov = 'treat',
                      mar_prop = c(0.1, 0.2, 0.4),
                      type = 'mar'),
  sample_size = 1000
)

simulate_fixed(data = NULL, sim_arguments) |>
  simulate_error(sim_arguments) |>
  generate_response(sim_arguments) |>
  generate_missing(sim_arguments)
```

The proportion of missing data for each level of the "treat" attribute is shown in Table 6.1. The sample size was increased to 10,000 to show that, in

TABLE 6.1 Proportion of missing data by each treatment level.

treat	Proportion Missing
Control	0.104
Low Dose	0.203
High Dose	0.398

expectation, the proportion of missing data is very close to the specified values in the `mar_prop` argument.

6.4.2.1 Numeric Missing Attribute

If the attribute specified within `miss_cov` is numeric, then a range of values can be specified up to the number of unique values to be simulated. For example, in the example used here, the age attribute can take on values between 30 and 60, resulting in a total of 31 unique values. Therefore, up to 31 unique proportion values can be specified where the first value specified would be associated with the missing data proportion for 30 years old, the second for 31 years old, and so on, up to the final proportion being for 60 years old.

If fewer than 31 unique proportions are defined, the missing data proportions are grouped together for a range of attributes. For example, if five missing data proportions are specified, the age attribute would be split into five age ranges of similar width. In this case, it would be about \$31/5 $\approx$ 6\$ year widths.

```
set.seed(321)

sim_arguments <- list(
  formula = y ~ 1 + age + treat,
  reg_weights = c(0.75, 0, 0.33),
  fixed = list(age = list(var_type = 'ordinal', levels = 30:60),
               treat = list(var_type = 'factor',
                            levels = c('Treatment', 'Control'))),
  missing_data = list(new_outcome = 'y_missing', miss_cov = 'age',
                      mar_prop = c(0.1, 0.2, 0.5),
                      type = 'mar'),
  sample_size = 1000
)

simulate_fixed(data = NULL, sim_arguments) |>
  simulate_error(sim_arguments) |>
  generate_response(sim_arguments) |>
  generate_missing(sim_arguments)
```

TABLE 6.2 Proportion of missing data by each age width.

Age Range	Proportion Missing
30, 40	0.100
41, 50	0.197
51, 60	0.514

Table 6.2 shows the missing proportions for the numeric attribute with age ranges associated with each `mar_prop` value. The sample size was again specified as 10,000 to show that, in expectation, the missing data values would be close to the specified values.

6.4.3 Dropout Missing Data

Dropout missing data is a type that can occur with longitudinal data in which an individual or unit being measured drops out of the study. For example, if an individual was coming to a hospital to measure their blood pressure every week for ten weeks. In this situation, it is plausible that some individuals would stop participating in the study and no longer have valid measurements. In these situations, this individual would dropout from the study. This type of missing data can be accommodated when using linear mixed models, and the number of measurement occasions for each individual can help inform model estimates.

The dropout missing data can be specified with the `type = 'dropout'` missing data argument. This type of missing data is only possible with longitudinal-type designs. Therefore, another argument representing the cluster variable must be specified with the argument `clust_var`. This is the name of the cluster argument specified as the "ID" within the `(random effect | ID)` portion of the model formula. The remaining arguments have been seen before, `miss_prop` represents the likelihood an individual will dropout from the study, and `new_outcome` is the name of the outcome in the simulated data that contains the missing data.

```
set.seed(321)

sim_arguments <- list(
  formula = y ~ 1 + time + treat + (1 + time| id),
  reg_weights = c(4, 0.5, 0.33),
  fixed = list(time = list(var_type = 'time'),
               treat = list(var_type = 'factor',
                            levels = c('Control', 'Treatment'),
                            var_level = 2)),
```

TABLE 6.3 Proportion of missing data for each measurement occasion (i.e., time).

time	Proportion Missing
0	0.000
1	0.005
2	0.040
3	0.070
4	0.085
5	0.135
6	0.170
7	0.205
8	0.240
9	0.240

```
  randomeffect = list(int_id = list(variance = 8, var_level = 2),
                      time_id = list(variance = 3, var_level = 2)),
  missing_data = list(miss_prop = .25, new_outcome = 'y_missing',
                      clust_var = 'id', type = 'dropout'),
  sample_size = list(level1 = 10, level2 = 20)
)

simulate_fixed(data = NULL, sim_arguments) |>
  simulate_randomeffect(sim_arguments) |>
  simulate_error(sim_arguments) |>
  generate_response(sim_arguments) |>
  generate_missing(sim_arguments)
```

Table 6.3 shows the proportion of missing data for each time point. Since the missing data is in dropout mode, the proportion of missing data increases as time increases. This occurs as each individual has a 20% likelihood of dropping out from the study; therefore, once they dropout, no subsequent measurements would be made. From the table, one individual dropped out after the first measurement, and they would be missing all subsequent data points.

Overall, the number of individuals with missing data would be 48 people. In the example above, 200 individuals were simulated, which results in a missing proportion of 0.24, close to the specified value of 0.2. Increasing the level two sample size would make this proportion closer to the specified value.

6.5 Correlated Fixed or Random Effects

Another way to make simulated data more realistic is to introduce a correlation between model elements. With data collected in the real world, attributes are correlated by nature. This means that the attributes have overlapping variance and may not mutually explain unique portions of variance. When estimating power, correlated attributes may also increase the size of the standard errors, which could impact power. Therefore, it can be important to explore how correlated attributes can empirically impact the standard errors and, as a result, power.

The **simglm** package allows for correlating the fixed attributes and random effects. It is possible to correlate either fixed attributes, random effects, or both simultaneously. Correlating these elements of the model is done through the "correlate" simulation argument. Examples of correlating only fixed effects, only random effects, and both will be shown in detail.

6.5.1 Correlated Fixed Effects

Correlating fixed effects are specified via the "fixed" named element of the `correlate` simulation argument. Inside the `fixed` element, a data frame is passed that specifies the pairwise correlations. Table 6.4 shows the structure that this data frame should take. Notice that there are two columns, generically named "x" and "y". These are where the names of the attributes are specified; these much match the model formula from the simulation arguments. The final column is named "corr" and represents the correlation between each pair of attributes. For example, the correlation between two attributes, "act" and "GPA," is specified to be 0.5 from Table 6.4.

The data frame is added to the `correlate` simulation argument in the code below. Notice the match between the attributes specified in the `correlate` simulation argument and the model formula specified first. Since three attributes are specified in the model formula, all attributes will be correlated.

TABLE 6.4 Structure for the fixed effect correlation specification.

x	y	corr
act	gpa	0.5
act	commute_time	0.6
gpa	commute_time	0.2

TABLE 6.5 Pearson correlations of the simulated fixed attributes.

term	commute_time	act	gpa
commute_time	NA	NA	NA
act	0.606	NA	NA
gpa	0.184	0.4861	NA

```
sim_arguments <- list(formula = y ~ 1 + act + gpa + commute_time,
                  fixed = list(act = list(var_type = 'ordinal',
                                      levels = 15:36),
                          gpa = list(var_type = 'continuous',
                                      mean = 2,
                                      sd = .5),
                          commute_time = list(var_type = 'continuous',
                                      mean = 15,
                                      sd = 6)),
                  correlate = list(
                    fixed = data.frame(x = c('act', 'act', 'gpa'),
                                    y = c('gpa', 'commute_time',
                                          'commute_time'),
                                    corr = c(0.5, 0.6, 0.2))),
                  sample_size = 1000)

simulate_fixed(data = NULL, sim_arguments) |>
  correlate_variables(sim_arguments)
```

Table 6.5 shows the correlations for the new attributes. These values are close to the correlations desired. The data were simulated with a sample size of 10,000; smaller sample sizes would likely have more variation in the estimated correlations but should converge to the population value with repeated simulations. Pearson correlations are reported for simplicity; for truly ordinal measures, rank-based or polychoric correlations may be more appropriate.

```
## Correlation computed with
## * Method: 'pearson'
## * Missing treated using: 'pairwise.complete.obs'
```

6.5.2 Correlate Random Effects

Correlating random effects is performed similarly to that of fixed effects. The only notable difference is the use of the `random` named argument to the `correlate` simulation argument instead of `fixed`. Below, an example that correlates the random intercepts and the random slopes is shown. A positive correlation is specified, meaning that those who tend to have higher random intercepts would also tend to have larger random slopes. The names specified within the x and y vectors of the data frame correspond to the names chosen inside the `randomeffect` simulation argument.

```
sim_arguments <- list(formula = y ~ 1 + act + gpa + sat + (1 + act | id),
                      fixed = list(act = list(var_type = 'continuous',
                                              mean = 20,
                                              sd = 4),
                                   gpa = list(var_type = 'continuous',
                                              mean = 2,
                                              sd = .5),
                                   sat = list(var_type = 'continuous',
                                              mean = 500,
                                              sd = 100)),

                      randomeffect = list(int_id = list(variance = 8, var_level = 2),
                                          act_id = list(variance = 3, var_level = 2)),
                      sample_size = list(level1 = 100, level2 = 5000),
                      correlate = list(random = data.frame(x = 'int_id', y = 'act_id',
                                                           corr = .3))
                      )

simulate_randomeffect(data = NULL, sim_arguments) |>
  correlate_variables(sim_arguments)
```

```
## Correlation computed with
## * Method: 'pearson'
## * Missing treated using: 'pairwise.complete.obs'
```

The estimated correlation from the data prior to correlating is 0.01, and the new correlation after correlating the random effects is 0.31.

6.5.3 Combine Fixed and Random Correlations

Combining the code from each section above is possible when both the fixed and random effects should be correlated. To specify both, a separate data frame that identifies the correlations for the fixed and random effects is added to the `correlate` simulation argument. The following code combines both examples discussed independently above. When both are specified, the `correlate_variables()` function will automatically process both correlations.

```
sim_args <- list(formula = y ~ 1 + act + gpa +
                   commute_time + (1 + act | id),
                 fixed = list(act = list(var_type = 'ordinal',
                                         levels = 15:36),
                              gpa = list(var_type = 'continuous',
                                         mean = 2,
                                         sd = .5),
                              commute_time = list(var_type = 'continuous',
                                                  mean = 15,
```

```
                                                 sd = 6)),
                  randomeffect = list(
                    int_id = list(variance = 8, var_level = 2),
                    act_id = list(variance = 3, var_level = 2)),
                  sample_size = list(level1 = 100, level2 = 5000),
                  correlate = list(
                    fixed = data.frame(x = c('act', 'act', 'gpa'),
                                       y = c('gpa', 'commute_time',
                                             'commute_time'),
                                       corr = c(0.5, 0.6, 0.2)),

                    random = data.frame(x = 'int_id', y = 'act_id',
                                        corr = .3)
                  )
)

simulate_fixed(data = NULL, sim_args) |>
  simulate_randomeffect(sim_args) |>
  correlate_variables(sim_args)
```

The simulation features introduced in this chapter provide the foundation for the simulation-based power analyses developed later in the book. Choices about effect size, variance structure, missingness, and correlation directly shape both the precision of estimates and the ability to detect effects, making these components central to understanding power in realistic study settings.

7

What is Power?

7.1 Chapter Overview

This chapter provides a conceptual foundation for statistical power, framing it within the null hypothesis significance testing framework commonly used in applied research. Power is defined in terms of hypothesis testing, effect detection, and long-run operating characteristics of statistical procedures. The chapter discusses the relationship between power, Type I and Type II error, sample size, and effect size, and clarifies common misconceptions about what power does and does not represent. By grounding power in statistical decision-making, the chapter prepares readers for later chapters that move beyond analytic power calculations. The discussion emphasizes interpretation and practical implications rather than mathematical derivations, making the concepts accessible to applied researchers.

Power is the ability to detect an effect of interest. Put into statistical terms; this could be phrased as detecting an effect size that is not zero or rejecting the statistical null hypothesis (or, more accurately, providing evidence that the statistical null hypothesis is unlikely to be true with the sample data). This type of power is typically framed within the null hypothesis significance testing (NHST) framework. The elements of this framework will be described briefly below, and the pros/cons of this approach will also be discussed.

When defining power, it is important first to define the statistical hypotheses that frame the question(s) of interest. Although these will likely vary depending on the data collection method and statistical model fitted, the null and alternative hypotheses will typically follow this general framework when the analysis can be framed as a linear regression:

$$H_0 : \beta_t = 0$$
$$H_A : \beta_t \neq 0$$

where H_0 is the null hypothesis representing that the regression parameter (β_t) is equal to 0 in the population. The null hypothesis is the statistical hypothesis with respect to which the p-value is interpreted in frequentist statistics. Stated another way, the p-value represents the probability of observing the obtained result, or one more extreme, assuming the null hypothesis is true. When

DOI: 10.1201/9781003177876-7

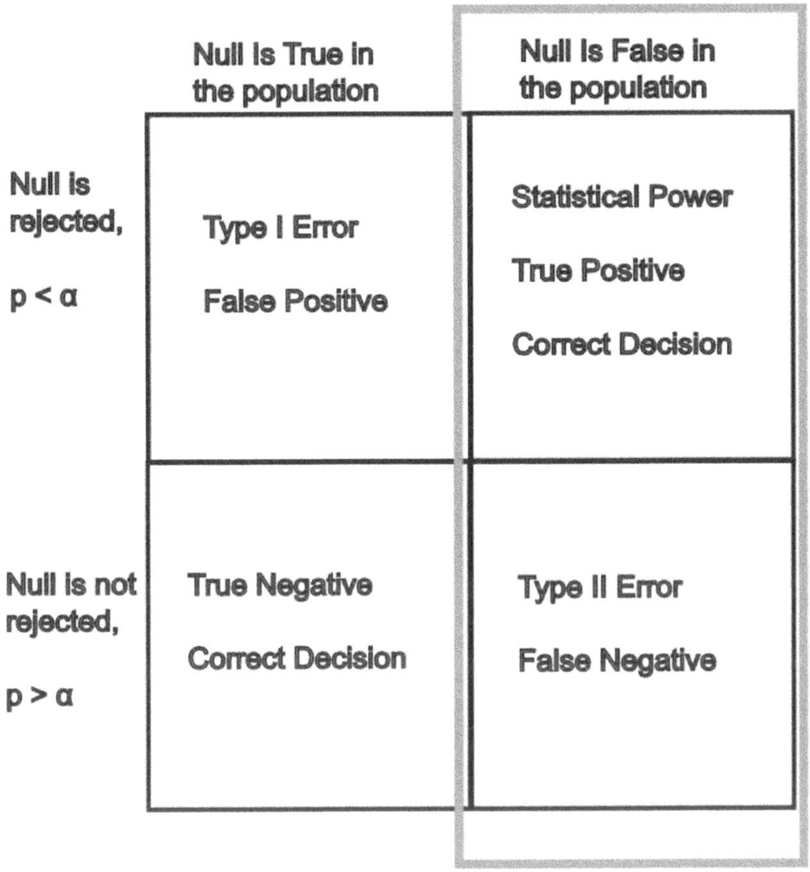

FIGURE 7.1 A decision diagram for frequentist hypothesis testing using an arbitrary cutoff.

evaluating power, the second hypothesis, H_A, or the alternative hypothesis, is of greater importance. This hypothesis represents the effect of interest or, more generally, the assumed data-generating distribution of the effect. In many power analyses, the alternative hypothesis corresponds to a nonzero parameter value in the population, a common (but not always appropriate) assumption when evaluating power. Statistical power is defined as the proportion of times the null hypothesis would be rejected, under a specified decision rule, when the effect size is not zero. These definitions rely on classical interpretations of the p-value, which will be discussed in more detail in the next section.

Figure 7.1 may be familiar from a prior statistics course. This table shows the possible decisions that can be made when using the hypothesis testing framework above. The leftmost column, labeled null, is true in the population and is less central to power calculations, though still relevant through the choice of α. The cells of that table's column are filled in for completeness but will not be expanded further here.

The second column, highlighted by a grey box, is more interesting when discussing power. In this column, the population effect size is known (or assumed) to not equal to zero. The rows in Figure 7.1 depict the statistical decision that is made when using the NHST framework. In the top row, the null hypothesis is rejected, with a p-value smaller than the specified α value. This would represent the correct decision, a true positive, and the case where statistical power can be estimated. The second row represents the situation where the null hypothesis is not rejected, with a p-value greater than the specified α value. This would represent a type II error or a false negative. The conclusion when in this cell of the figure would be to conclude that the effect size is not different from zero, when it really is in the population. Given that we know or are assuming that the effect size is not zero in the population when in this column of the figure, this decision would most likely occur due to a lack of statistical precision, and an increase in sample size is one way to increase this precision.

This classic power diagram dichotomizes the continuous p-value into one larger or smaller than the specified α value. As is the case when dichotomizing a continuous data attribute (Royston et al., 2006), there is a loss of information when dichotomizing the p-value into larger or smaller than α. There has been much discussion on why the p-value should not be interpreted relative to fixed, researcher-specified thresholds such as 0.05 (Goodman, 2008). A primary concern is that these thresholds are conventional rather than principled and do not arise from the data-generating process, the scientific question, or the consequences of decision errors. Dichotomizing results into "significant" and "not significant" based on an arbitrary cutoff can lead to exaggerated differences between results that are practically indistinguishable (e.g., $p = 0.049$ versus $p = 0.051$), while obscuring the underlying uncertainty in the estimated effect. This practice encourages binary decision-making and overconfidence in results that barely cross a threshold, while discounting evidence that may be substantively meaningful but fails to meet the chosen cutoff. As a result, threshold-based interpretations of p-values can distort scientific reasoning, incentivize selective reporting, and shift attention away from effect sizes, uncertainty, and the plausibility of the underlying assumptions.

It is possible to frame the traditional power estimation more directly regarding precision. For example, instead of focusing on p-values, it is possible to estimate a standard error for the effect size and compute a desired confidence interval to determine if it contains practically relevant values. However, this still creates many data conditions (i.e., normally distributed errors, homogeneity of variance, equal group sizes, etc.), which may not be practical for real-world data.

Furthermore, the confidence interval, although giving information about the likely range of the effect size, is related to and often interpreted similarly to the p-value. That is, the confidence interval is often interpreted as not containing zero or the null effect. Therefore, the effect size is then meaningful (Feinstein, 1998). This will be a recurring theme moving forward as the articulation for choosing power by simulation and the alternative framework posed in Chapter @ref{alt-power-hypoth}.

Relatedly, Wasserstein and Lazar (2016) provides a statement by the American Statistical Association on p-values. The statement provides six principles for which the p-value should or should not be interpreted. Additional supplemental articles provide even further detail; for example, Greenland et al. (2016) discusses common misinterpretations for the p-values, confidence intervals, and statistical power in more depth.

7.2 Why Do We Care About Power?

Statistical power is typically used for study planning and is often required by grant funding agencies. As the title suggests, this book advocates for estimating power by simulation, but there are other frameworks for estimating power. Depending on the method, closed-form mathematical equations may provide an alternative to estimating power by simulation. Closed-form solutions can be a helpful guide when data meet the required assumptions. For example, most closed-form solutions assume the data are approximately Normal, do not have heterogeneity of variance, and have equal sample sizes across groups, among other data conditions. These may not be realistic for a given statistical design.

When estimating power via closed-form solutions, statistical software is dedicated to this problem; for example, G*Power (Faul et al., 2007), PowerUp! (Dong et al., 2015), or Optimal Design (Raudenbush et al., 2011). For many common statistical methods, such as t-tests or ANOVA, general-purpose software systems also have routines for estimating power. For example, within R, `power.t.test()` will estimate power for one and two sample t-tests. It is important to know the data conditions made when using this software; for example, the `power.t.test()` function will make the data condition of homogeneity of variance, equal group sizes, and normally distributed error terms in the two sample t-test case.

7.3 A Priori vs Post Hoc Power

When estimating power, two frameworks are often discussed: a priori or post hoc power. A priori power is commonly used for study planning to help determine how much data to collect and the sample size. On the other hand, post hoc power is commonly used to estimate power after the study is completed. Post hoc power is typically used to justify why a study did not find the predefined cutoff for "significance." The difference between post hoc and a priori power has been written about extensively in the statistical literature (Hoenig and Heisey, 2001; Levine and Ensom, 2001; O'Keefe, 2007), but the consensus is that post hoc power is rarely a useful quantity to estimate. Hoenig and Heisey (2001) discusses two common situations where post hoc power can sometimes be suggested for a nonrejected null hypothsis, for calculating observed power or detectable effect sizes. Hoenig and Heisey (2001) also defines the term, power approach paradox, for a common misinterpretation of post-hoc power estimates and to why some suggest they are useful. Hoenig and Heisey (2001) define the power approach paradox where high observed power does not provide stronger evidence for a null hypothesis that is not rejected. Put another way, observed power adds no information beyond the p-value itself. As a result, this book will be framed around a priori power and study planning rather than estimating power after a study is completed. As such, a priori power will be referred to simply as power or statistical power.

Statistical power is best completed well in advance of any data collection. Statistical power, when conducted a priori, should be considered a descriptive analysis where the impact of different data conditions would be explored to understand how sensitive power estimates are to the data being collected. For example, how much could power vary if heterogeneity of variance is 10 to 1 instead of 3 to 1, or what happens if 15% of the data are missing compared to 5%? These are questions that can be explored when estimating power from a simulation framework more easily than using closed-form equations that make data conditions (i.e., no missing data, normally distributed residuals, etc.) to estimate power.

7.4 Minimal Detectable Effect Size

More recently, power has been framed as the smallest effect size that a statistical analysis could successfully detect about 80% of the time (i.e., with a fixed statistical power of 0.80). This is sometimes referred to as the minimal detectable effect size (MDE) (Bloom, 1995). Although the terminology differs, the framework for estimating power is typically the same, except instead of estimating power, the effect size that meets a fixed power threshold is estimated. This framework still resolves around the null hypothesis significance testing. The inference around MDE is that given the effect size, what sample size would be needed to reject the null hypothesis 80% of the time? The mechanics around null hypothesis significance testing apply here. For example, researchers would need to specify an α value, which reflects the threshold of a "significant" result (i.e., a p-value smaller than the α value is deemed "significant") or the likelihood of making a type I error. However, conditional on the assumed nonzero effect size, Type I error is no longer the relevant quantity for the simulation scenario as the effect size is assumed different from zero in the population. Instead, the MDE identifies how often, over repeated sampling, sufficient precision exists to estimate an effect that would not contain the null effect of zero. This helps to identify if the sampling plan contains enough data to estimate the effect size with some level of precision. The researcher can also specify if the significance test will be one or two-tailed. This is often two-tailed, but estimating an MDE for a one-tailed test is possible. The one-tailed test would give the researcher greater "power" because the statistical test is pooled to focus on one tail of the sampling distribution.

7.4.1 Selecting a Power Threshold

When selecting a power threshold, there is a trade-off between being more certain (i.e., having more statistical precision) in estimating the effect size versus needing more data. Collecting more data can be expensive and take significant amounts of time. Therefore, some level needs to be specified that balances the statistical precision needed compared to the time and money to collect more data.

The power threshold often specified is 80%, although, much like the alpha or type I error rate in classical statistics, 80% statistical power is a threshold that the researcher can specify. Although the researcher can specify this, when creating a power analysis for grant applications, justification of why the specific threshold was selected should be made. For example, grant reviewers would likely strongly question why a power threshold was significantly reduced but would likely not question more stringent power thresholds than are typical. Strategies to successfully defend the changing of the power threshold will depend on the specific project. However, strategies such as discussing the cost-benefit of collecting more data, the experimental nature of the research, small population sizes, or other characteristics can help frame why a specific power threshold was chosen.

7.5 Estimating Power by Simulation

Statistical power estimated via simulation-based methods is an alternative framework to the closed-form solutions discussed above. As discussed in Chapter 1, the primary benefits of power by simulation include:

- no need to depend on data conditions
- can use the same framework no matter the study design
- may be more accurate for the data being collected
- can study the impact of different data conditions.

These are not trivial components when estimating power, as some can strongly impact the estimated power. For example, heterogeneity in the treatment effect can substantially impact the estimated power. In this scenario, closed-form solutions would typically overestimate power when strong heterogeneity in the treatment effect is found in the collected study. Another factor with strong implications for estimated power is that violations of modeling assumptions often require substantially more data to achieve the same level of precision.

8

Power by Simulation

8.1 Chapter Overview

This chapter introduces simulation-based power analysis as a flexible alternative to traditional analytic power calculations. The chapter outlines the basic workflow of simulation-based power analysis, including data generation, model fitting, hypothesis testing, and result aggregation. Practical examples demonstrate how simulation can accommodate complex designs, non-standard models, and realistic data conditions that are difficult or impossible to address analytically. The chapter emphasizes reproducibility and transparency, showing how simulation results can be summarized, visualized, and interpreted. By framing power as an empirical property of an analysis under known conditions, the chapter highlights the strengths of simulation for answering applied design questions.

The **simglm** package has built helper functions to support power analysis so that users do not need to create their own functions to perform the analysis. These include functions for fitting the models to the simulated data, `model_fit()`, extracting model coefficients `extract_coefficients()`, replicating the simulation `replicate_simulation()`, and computing power statistics `compute_statistics()`. Each of these will be discussed in more detail in the following sections along with any new simulation arguments that may be needed for each of them.

8.2 Model Fitting

The function, `model_fit()`, is one that fits a specified model to the simulated data. If no simulation arguments are specified for the `model_fit()` function, the default behavior is as follows:

- fits the same model as specified in the `formula` simulation arguments
- matches the type of data, `lm()` for non-nested data, `lmer()` from the **lme4** package (Bates et al., 2015) for nested data

DOI: 10.1201/9781003177876-8

The following is an example of this behavior, using the `model_fit()` function with no specified simulation arguments.

```
set.seed(321)

sim_arguments <- list(
  formula = y ~ 1 + weight + age + sex,
  fixed = list(weight = list(var_type = 'continuous',
                             mean = 180, sd = 30),
               age = list(var_type = 'ordinal', levels = 30:60),
               sex = list(var_type = 'factor',
                          levels = c('male', 'female'))),
  error = list(variance = 25),
  sample_size = 10,
  reg_weights = c(2, 0.3, -0.1, 0.5)
)

simulate_fixed(data = NULL, sim_arguments) |>
  simulate_error(sim_arguments) |>
  generate_response(sim_arguments) |>
  model_fit(sim_arguments)
```

The output from the `model_fit()` function is a model object. By itself, the `model_fit()` function isn't the most useful, rather, it is much more useful when combined with the `extract_coefficients()` function.

8.3 Extract Coefficients

To continue in the model pipeline, after fitting a model object, it is often of interest to extract the estimated parameters. The `extract_coefficients()` function in the **simglm** package aids the user in extracting fixed effect coefficients from the fitted model. The `extract_coefficients()` function does not take any arguments, rather, it detects the type of model object and extracts the estimated model parameters. It returns a tidy data frame of fixed-effect estimates (estimate/SE/test statistic/p-value when available), using broom-style tidying using the **broom** package (Robinson et al., 2022) or internally.

```
set.seed(321)

sim_arguments <- list(
  formula = y ~ 1 + weight + age + sex,
```

TABLE 8.1 Extracted coefficients from simulated data with the extract coefficients function.

term	estimate	std.error	statistic	p.value
(Intercept)	2.06	0.40	5.17	0
weight	0.30	0.00	178.16	0
age	−0.10	0.01	−17.13	0
sexfemale	0.41	0.10	4.10	0

```
  fixed = list(weight = list(var_type = 'continuous',
                             mean = 180, sd = 30),
               age = list(var_type = 'ordinal', levels = 30:60),
               sex = list(var_type = 'factor',
                          levels = c('male', 'female'))),
  error = list(variance = 25),
  sample_size = 10000,
  reg_weights = c(2, 0.3, -0.1, 0.5)
)

simulate_fixed(data = NULL, sim_arguments) |>
  simulate_error(sim_arguments) |>
  generate_response(sim_arguments) |>
  model_fit(sim_arguments) |>
  extract_coefficients()
```

Table 8.1 shows the output from the `extract_coefficients()` function. This function pulls information that is commonly extracted via the `summary()` function, but the information here is extracted into a data frame. This allows for the objects to be combined more readily when multiple replications are performed.

8.4 Optional Model Fit Arguments

There are a few model fit arguments that are often of interest when fitting models using the `model_fit()` function that may be of most interest. First, fitting a simplified model compared to the simulated population model. Secondly, it could be of interest to fit a model through of different modeling framework. These are possible to specify through optional arguments specified inside the `model_fit` simulation arguments. Each of these topics will be discussed in more detail in the next couple sections.

8.4.1 Simplified Model Fitting

Fitting a simplified model could be of interest if there is a parameter that is known to be important in modeling a specific outcome, but for some reason this attribute is unable to be collected. In this situation, the population model would want to include this effect, but the model fitting would omit this effect. The misspecified model can have larger standard errors and can induce bias in remaining coefficients if the omitted predictor is correlated with included predictors.

This can be specified via a `formula` named element inside the `model_fit` simulation arguments. Similar to the formula specified for the population simulated data, the formula would contain a subset of terms from the simulated data. In the example shown below, the population model included an intercept and three other terms (i.e., weight, age, and sex). However, the model fitting formula omits the weight effect and only models the intercept, age, and sex effects.

```
set.seed(321)

sim_arguments <- list(
  formula = y ~ 1 + weight + age + sex,
  fixed = list(weight = list(var_type = 'continuous',
                             mean = 180, sd = 30),
               age = list(var_type = 'ordinal', levels = 30:60),
               sex = list(var_type = 'factor',
                          levels = c('male', 'female'))),
  error = list(variance = 25),
  sample_size = 10000,
  reg_weights = c(2, 0.3, -0.1, 0.5),
  model_fit = list(formula = y ~ 1 + age + sex)
)

simulate_fixed(data = NULL, sim_arguments) |>
  simulate_error(sim_arguments) |>
  generate_response(sim_arguments) |>
  model_fit(sim_arguments) |>
  extract_coefficients()
```

Table 8.2 shows the extracted coefficients from the simplified model. You can notice that the weight effect is missing and the increase in standard errors compared to the extracted coefficients from the full population model in 8.1. You can also see that due to the misspecified model, the interpretation of the intercept changes (due to the scale of the weight attribute not containing 0 in the first model), but the estimated effects for the other terms are similar.

TABLE 8.2 Extracted coefficients from simulated data with a simplified model fitted.

term	estimate	std.error	statistic	p.value
(Intercept)	55.60	0.53	104.10	0.00
age	−0.09	0.01	−7.74	0.00
sexfemale	0.38	0.20	1.88	0.06

8.4.2 Different Model Framework

Another optional argument that may be particularly interesting, is the ability to change the model fitting framework. For example, the linear mixed model could be the generating model, however, it may be of interest to fit a generalized estimating equations model to compare how much, if at all, model estimates or standard errors differ between the two approaches.

The following example simulates longitudinal data with 10 time points and a single treatment attribute. The simulated data also has two random effects for each person, a random intercept and random slope. Finally, the outcome was specified as binary inside the simulation argument, "outcome_type". Often these type of data are modeled with a generalized linear mixed model, one such way to model this is with the `glmer()` function from the **lme4** package (Bates et al., 2015). The code below fits the model this way, where the family argument is specified with `family = binomial`.

```
set.seed(321)

sim_arguments <- list(
  formula = y ~ 1 + time + treat + (1 + time| id),
  fixed = list(time = list(var_type = 'time',
                           time_levels = c(0, 0.5, 1, 1.5, 2,
                                           2.5, 3, 4, 5, 6)),
               treat = list(var_type = 'factor',
                            levels = c('Treatment', 'Control'),
                            var_level = 2)),
  reg_weights = c(0.4, 0.2, -0.5),
  randomeffect = list(int_id = list(variance = 8, var_level = 2),
                 time_id = list(variance = 3, var_level = 2)),
  error = list(variance = 5),
  outcome_type = 'binary',
  sample_size = list(level1 = 10, level2 = 100),
   model_fit = list(
    model_function = lme4::glmer,
    family = binomial
```

TABLE 8.3 Estimated fixed effects for longitudinal data simulation using generalized linear mixed models.

term	estimate	std.error	statistic
(Intercept)	−0.0881	5e-04	−173.37
time	0.3155	5e-04	619.92
treatControl	0.0064	5e-04	12.59

```
  )
)
```

```
simulate_fixed(data = NULL, sim_arguments) |>
  simulate_error(sim_arguments) |>
  simulate_randomeffect(sim_arguments) |>
  generate_response(sim_arguments) |>
  model_fit(sim_arguments) |>
  extract_coefficients()
```

Table 8.3 shows the estimated fixed effect parameters from this model. One can see that the standard error for the treatment effect, likely the one that is of most interest in this type of model, is the one that has a larger standard error indicating much more uncertainty in this estimate. This is likely due to only 100 individuals being in the study.

An alternative modeling framework for data with a nested structure is the use of generalized estimating equations (GEE) (Liang and Zeger, 1986; Diggle, 2002). Unlike mixed-effects models, GEE does not include random effects; instead, it produces population-averaged estimates while accounting for within-cluster dependence through a specified working correlation matrix.

In the `model_fit` simulation arguments below, the `geeglm()` function from the geepack R package is used (Halekoh et al., 2006). As before, the family is specified as binomial; however, additional arguments unique to GEE models are required. These include the "id" argument, which identifies the clustering variable, and the "corstr" argument, which specifies the assumed working correlation structure. See the `geeglm()` documentation for further details. Finally, because GEE models do not include random effects, the model formula must be modified to include only fixed effects.

```
set.seed(321)

sim_arguments <- list(
```

```
  formula = y ~ 1 + time + treat + (1 + time| id),
  fixed = list(time = list(var_type = 'time',
                           time_levels = c(0, 0.5, 1, 1.5, 2,
                                           2.5, 3, 4, 5, 6)),
               treat = list(var_type = 'factor',
                            levels = c('Treatment', 'Control'),
                            var_level = 2)),
  reg_weights = c(0.4, 0.2, -0.5),
  randomeffect = list(int_id = list(variance = 8, var_level = 2),
                 time_id = list(variance = 3, var_level = 2)),
  error = list(variance = 5),
  outcome_type = 'binary',
  sample_size = list(level1 = 10, level2 = 100),
  model_fit = list(
    model_function = geepack::geeglm,
    formula = y ~ 1 + time + treat,
    id = 'id',
    family = binomial,
    corstr = 'ar1'
  )
)

simulate_fixed(data = NULL, sim_arguments) |>
  simulate_error(sim_arguments) |>
  simulate_randomeffect(sim_arguments) |>
  generate_response(sim_arguments) |>
  model_fit(sim_arguments) |>
  extract_coefficients()
```

Table 8.4 shows the estimated GEE model results. In this scenario, the estimated treatment effect, which is likely of most interest, is estimated with smaller standard errors compared to the generalized linear mixed model. This is not surprising given the difference in interpretation between these two modeling frameworks: GEE models yield population-averaged (marginal) effects,

TABLE 8.4 Estimated model parameters when a GEE model was fitted.

term	estimate	std.error	statistic	p.value
(Intercept)	−0.03	0.13	0.07	0.80
time	0.08	0.03	5.16	0.02
treatControl	0.20	0.13	2.56	0.11

whereas generalized linear mixed models yield conditional effects, interpreted given the random effects, typically evaluated at zero (Diggle, 2002; Zeger et al., 1988).

8.5 Replicating Power Simulation

The **simglm** R package contains functionality to replicate power analyses to include more than one replication. This is fundamental to get stable estimates of power for a specific condition. When conducting power by simulation, these are estimates and replicating the data generating conditions many times can produce more stable power estimates. This is related to the errors found in Monte Carlo simulation, more details on the errors when simulating data can be found in Morris et al. (2019) and shown in Chapter 10.

The `replicate_simulation()` function is used for this purpose. This function works a bit differently than other functions in the book in that it uses a wrapper type function, `simglm()`. This function looks at the simulation arguments and creates a pipeline of functions that it should used based on the arguments specified. For example, to generate a mixed model with random effects and the `simulate_randomeffect()` function, the `randomeffect` simulation argument needs to be specified, otherwise this step is skipped in the data simulation pipeline. This would also be true for the "error" simulation argument. If the `error` argument is not specified, the `simulate_error()` simulation function would not be processed.

8.5.1 Power Arguments for Replicating Simulation

Specifically for power estimation, the three additional simulation arguments that should be specified would include, `replications`, `model_fit`, and `extract_coefficients`. The `model_fit` argument specifies which arguments to pass to the model fitting procedure and explicitly call the `model_fit()` function. At a minimum, the formula to be fitted should be specified, but other optional arguments that were discussed in Section 8.4 are possible to be specified here. These would include different model fitting functions, different model fitting formulas, and any other arguments needed to pass to the model fitting function.

Finally, the `extract_coefficients` simulation argument should be specified as `TRUE` if the model fixed effects should be extracted from the model that was fitted to the simulated data. When this is specified as `TRUE`, then the `extract_coefficients()` function is added to the chain of functions called in the data processing.

The below example shows this pipeline for a linear regression model. With the simulation arguments specified in the code, the function pipeline would be:

1. `simulate_fixed()`
2. `simulate_error()`
3. `generate_response()`
4. `model_fit()`
5. `extract_coefficients()`

```
set.seed(321)

sim_arguments <- list(
  formula = y ~ 1 + weight + age,
  fixed = list(weight = list(var_type = 'continuous',
                             mean = 180, sd = 30),
               age = list(var_type = 'ordinal', levels = 30:60)),
  error = list(variance = 25),
  sample_size = 10,
  reg_weights = c(2, 0.3, -0.1),
  replications = 2,
  model_fit = list(formula = y ~ 1 + weight + age),
  extract_coefficients = TRUE
)

replicate_simulation(sim_arguments)
```

```
## [[1]]
## # A tibble: 3 x 5
##   term        estimate std.error statistic  p.value
##   <chr>          <dbl>     <dbl>     <dbl>    <dbl>
## 1 (Intercept) -1.39       10.4     -0.133   0.898
## 2 weight       0.294       0.0418   7.04    0.000205
## 3 age         -0.000817    0.176   -0.00464 0.996
##
## [[2]]
## # A tibble: 3 x 5
##   term        estimate std.error statistic p.value
##   <chr>          <dbl>     <dbl>     <dbl>   <dbl>
## 1 (Intercept)  -15.2      25.5      -0.595  0.571
## 2 weight         0.330     0.0949    3.48   0.0102
## 3 age            0.139     0.249     0.558  0.594
```

8.5.2 Compute Power Statistics

Once the `replicate_simulation()` function is called, there is a helper function that computes some commonly reported power statistics, `compute_statistics()`. This function will take the model estimates across the simulated data and aggregate over the number of replications. The default behavior for the `compute_statistics()` function is to compute three types of summary statistics:

1. Power statistics: average parameter estimates, power estimate, average test statistic, critical value used for power.
2. Type I Error Rates: type I error, average adjusted test statistic, critical value for type I error.
3. Precision: Standard deviation of the parameter estimates, average standard error, and the precision ratio (standard deviation of parameter estimates / average standard error).

Before moving to the example, it should be noted that the type I error rates are computed using an adjusted test statistics as follows:

$$Adjusted\ \ Test\ \ Statistic = \frac{\hat{\beta} - \beta}{SE}$$

where $\hat{\beta}$ is an estimated regression coefficient, β is the population regression coefficient specified via the simulation arguments, and SE represents the estimated standard error. This has the effect of centering the test statistic around the population value, a case where exploring the type I error would be appropriate. Therefore, with the **simglm** package, the type I error rates can be valid no matter what the generating model or population coefficients are. Using this type of statistic can be helpful when evaluating statistical methodology performance in different data situations.

Table 8.5 shows these statistics that are computed for the user by default. The power statistics show up first, then the type I error rates, then the precision statistics. These statistics are computed within each of the terms including in the fitted model. In the current example, there were three terms, the intercept,

TABLE 8.5 Computed power statistics when using the compute statistics function.

Term	Avg. Est.	Power	Param Est. SD.	Avg. SE	Prec. Ratio	Replications
(Intercept)	1.772	0.026	15.956	14.996	1.064	1000
age	−0.097	0.039	0.223	0.207	1.078	1000
weight	0.300	0.882	0.068	0.065	1.056	1000

the weight coefficient, and the age coefficient. As such, there are three rows of data when the `compute_statistics()` function is run.

```
set.seed(321)

sim_arguments <- list(
  formula = y ~ 1 + weight + age,
  fixed = list(weight = list(var_type = 'continuous',
                             mean = 180, sd = 30),
               age = list(var_type = 'ordinal', levels = 30:60)),
  error = list(variance = 25),
  sample_size = 10,
  reg_weights = c(2, 0.3, -0.1),
  replications = 1000,
  model_fit = list(formula = y ~ 1 + weight + age),
  extract_coefficients = TRUE
)

replicate_simulation(sim_arguments) |>
  compute_statistics(sim_arguments, alternative_power = FALSE)
```

If only a subset of the three types of statistics are desired, each can be controlled by three arguments within the `compute_statistics()` function. These boolean arguments are:

- `power`
- `type_1_error`
- `precision`

By default, all of these are set to `TRUE` which means the statistics corresponding to each are returned. However, suppose only the power and precision statistics are desired, then, `type_1_error = FALSE` can be set. This will omit the type I error statistics being computed and returned. Below, a more full example of the code is shown without the output.

```
replicate_simulation(sim_arguments) |>
  compute_statistics(sim_arguments,
                     type_1_error = FALSE, alternative_power = FALSE)
```

8.5.3 Optional Power Arguments

Although not required to generate power statistics, there are optional arguments that can be passed to the `compute_statistics()` function. These can include changing the α value, one or two tailed hypotheses, changing the distribution for the sampling distribution, or other parameters to pass to the sampling

TABLE 8.6 Computed power statistics when using the compute statistics function using the t-distribution.

Term	Avg. Est.	Power	Param Est. SD.	Avg. SE	Prec. Ratio	Replications
(Intercept)	1.772	0.018	15.956	14.996	1.064	1000
age	−0.097	0.036	0.223	0.207	1.078	1000
weight	0.300	0.854	0.068	0.065	1.056	1000

distribution function. For example, it is often of interest to use a t-distribution rather than a Normal distribution when sample sizes are small. It is possible to do this by specifying the quantile distribution function in R to use to generate the critical value.

The power arguments are specified within the simulation arguments with a named element called `power`. Within the power argument, the following are possible elements to include:

- `dist` is a quantile distribution function from R (a function that starts with "q", e.g., `qt()`). Default quantile distribution function is `qnorm()`.
- `alpha` is the α value, default value is $\alpha = .05$.
- `direction` specifies whether a one-sided or two-sided is used. The default is a two-sided hypothesis. One sided hypotheses can be specified as either "upper" or "lower".
- `opts` any additional arguments needed for the quantile distribution function specified. No default values.

Below, the power arguments are specified for estimating power with a t-distribution (i.e., using the `qt()` t-distribution quantile function), an $\alpha = .01$, one-sided upper tail hypothesis, and a t-distribution with eight degrees of freedom. Table 8.6 shows the computed statistics. Notice that the critical value for the power evaluation is now `round(qt(.01, 8, lower.tail = FALSE), 2)`, which is larger than the critical value found in the first example when a standard Normal distribution was used. In practice, choose degrees of freedom consistent with your test (e.g., residual degrees of freedom for OLS; for mixed models degrees of freedom can be more complex).

```
power = list(
  dist = 'qt',
  alpha = .01,
  direction = 'upper',
  opts = list(df = 8)
)
```

8.5.4 Vary Power Arguments Across Parameters

In the previous section, the power arguments were specified as one value, however, the **simglm** package is flexible to allow for the power arguments to be different for each regression parameter. For example, the previous example had three possible regression parameters for which power could be computed, the intercept, the weight, and the age parameters. The power arguments can be specified as a vector the same length as the number of regression parameters. The vector aligns with the order of the parameters specified in the formula simulation argument. This allows for the user to change the power argument for each specific regression parameter.

For example, perhaps the weight parameter was known to be a positive association so a one-sided upper tail test would be desired, but the age parameter was known to be a negative association so a one-sided lower tail test would be desired. This can be achieved by passing a vector of direction arguments instead of a single argument. The revised power simulation arguments would look as follows.

```
power = list(
  dist = 'qt',
  alpha = .01,
  direction = c('two-sided', 'upper', 'lower'),
  opts = list(df = 8)
)
```

Table 8.7 shows the updated power table. You can see the different hypothesis test direction by looking at the critical value column. First, the intercept row has a larger critical value, which is due to considering both tails. Secondly, you can see the critical value is smaller for the age and weight terms, is the same magnitude, but one is negative while the other is positive.

It is also possible to vary other arguments, but these are likely not as desirable as the direction argument. For example, one could use a different theoretical distribution or alpha value for all of the regression parameters as well. For example, the following updated power arguments code is possible.

```
power = list(
  dist = c('qt', 'qt', 'qt'),
  alpha = c(.05, .01, .001),
```

TABLE 8.7 Computed power statistics when using the compute statistics function while varying the direction argument for each regression parameter.

Term	Avg. Est.	Power	Param Est. SD.	Avg. SE	Prec. Ratio	Replications
(Intercept)	1.772	0.003	15.956	14.996	1.064	1000
age	−0.097	0.036	0.223	0.207	1.078	1000
weight	0.300	0.854	0.068	0.065	1.056	1000

TABLE 8.8 Computed power statistics when using the compute statistics function while varying all power arguments for each regression parameter.

Term	Avg. Est.	Power	Param Est. SD.	Avg. SE	Prec. Ratio	Replications
(Intercept)	1.772	0.026	15.956	14.996	1.064	1000
age	−0.097	0.036	0.223	0.207	1.078	1000
weight	0.300	0.539	0.068	0.065	1.056	1000

```
  direction = c('two-sided', 'upper', 'lower'),
  opts = list(df = 8)
)
```

Table 8.8 shows the power output for this analysis varying more than just the direction argument. In this case, you can see that the critical value varies in magnitude associated with the alpha value. More specifically, this means that a smaller alpha value leads to a larger critical value.

8.6 Note on Parallel Processing

One last topic to consider when estimating power, is speed. Estimating power can take time, particularly when a number of conditions are varied to understand the impact on power (see Chapter 9 for details on this). The **simglm** package uses the **future** package (Bengtsson, 2021) and ecosystem for parallel processing to spread out the computations to multiple cores or threads on the computer processor.

When using the future framework for parallel processing, a future plan needs to be set by the user. For most users desiring parallel processing to speed up computations, using `plan(multisession)` or `plan(multicore)` would cover most situations. Note, using `plan(multisession)` is currently recommended for those running computations within RStudio, but see documentation from the **future** package for recommendations.

To get a sense of the speedup that is possible, the power computations in the final code chunk in this chapter were performed with 10,000 replications using sequential (not parallel) compared to a multisession plan. The time to do these computations on an M1 MacBook Air (2021) inside RStudio were 58.2 seconds for sequential processing and was 25.8 seconds using the multisession parallel processing. This is more than a 50% reduction in processing time! Speedups vary by machine, number of workers, and model complexity; parallel overhead can dominate for small simulations.

9

Varying Arguments to Estimate Power

9.1 Chapter Overview

This chapter extends simulation-based power analysis by incorporating systematic variation of design parameters. Rather than estimating power for a single scenario, the chapter demonstrates how power surfaces can be explored across combinations of effect sizes, sample sizes, and model assumptions. The chapter discusses strategies for organizing simulation results and interpreting multidimensional summaries. Examples illustrate how varying parameters can reveal regions of adequate power, diminishing returns, and design tradeoffs. This approach encourages applied researchers to think of power as a function of design choices rather than a single numeric target.

Varying arguments to estimate power are similar to varying arguments for simulating data discussed in Chapter 5. The `vary_arguments` element of the simulation arguments is used to vary simulation or power arguments. The `vary_arguments` is a list that would include named simulation arguments. For example, if the sample size varied to different values, the `sample_size` simulation argument would be specified within the `vary_arguments` instead.

When specifying multiple simulation arguments inside the `vary_arguments` element, the default behavior is to cross all elements specified fully. For example, the code below specifies three different sample size arguments (i.e., 10, 50, and 150) and three different error variance arguments (i.e., 15, 25, and 50). As such, these levels would be fully crossed via a factorial design. With a factorial design, each sample size specified would be crossed for each error variance specified. Procedurally, when thinking about this for the simulation and power analysis framework, this would mean that data would be simulated for each crossed pair of sample size and error variance arguments. Table 9.1 shows what this would look like for this type of design.

When estimating power is the goal of the simulation, a few additional simulation arguments need to be specified, as discussed within Chapter 8. These include the `model_fit`, `extract_coefficients`, and optionally, but recommended, the `power` simulation arguments. These arguments are needed to fit an appropriate model, extract the coefficients, and pass any optional power arguments when computing the estimated power. The code chunk below adds the sample size

DOI: 10.1201/9781003177876-9

TABLE 9.1 Crossed factors specified within vary arguments.

sample_size	error
10	15
50	15
150	15
10	25
50	25
150	25
10	50
50	50
150	50

and error variance as varying arguments. The `replicate_simulation()` and `compute_statistics()` functions are used to replicate the simulation and power analysis across the nine varied simulation conditions (i.e., for each row in Table 9.1).

```
set.seed(321)

sim_arguments <- list(
  formula = y ~ 1 + weight + age,
  fixed = list(weight = list(var_type = 'continuous',
                             mean = 180, sd = 30),
               age = list(var_type = 'ordinal', levels = 30:60)),
  reg_weights = c(2, 0.3, -0.1),
  replications = 5000,
  model_fit = list(formula = y ~ 1 + weight + age),
  extract_coefficients = TRUE,
  power = list(
    dist = 'qt',
    alpha = .01,
    direction = 'upper',
    opts = list(df = 8)
  ),
  vary_arguments = list(
    sample_size = c(10, 50, 150),
    error = list(
      list(variance = 15),
      list(variance = 25),
```

```
      list(variance = 50)
    )
  )
)

replicate_simulation(sim_arguments) |>
  compute_statistics(sim_arguments, type_1_error = FALSE)
```

Figure 9.1 shows the output for the simulated data and estimated power for each term, sample size, and error variance. Notice that the y-axis was set to be unique for each panel in the figure to show the estimated power for each effect more clearly. All combinations' estimated power for the intercept and age term were small. The estimated power was higher for the weight term, indicating that when the sample size is greater than 10, the estimated power was close to 1 for all combinations. Functionally, this means that across the 5,000 replications, all p-values were smaller than the threshold specified (i.e., $\alpha = 0.01$). This would not mean that the power would be 1, but the analysis is limited to the 5,000 replications performed.

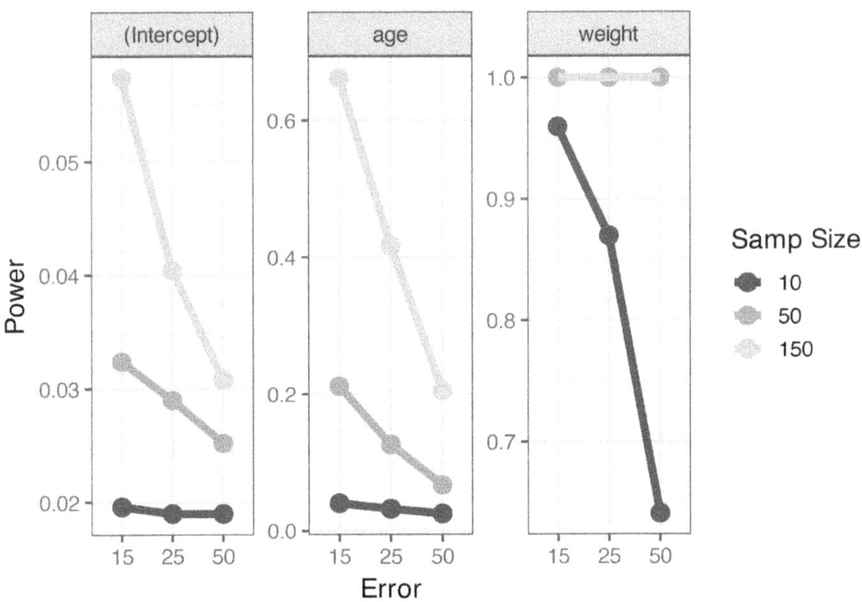

FIGURE 9.1 Estimated power by parameter, sample size, and error variance.

9.2 Varying Additional Arguments

The **simglm** package offers additional functionality to vary arguments other than those used to simulate data. For example, it may be interesting to estimate power for different fitted models specified via the `model_fit` arguments or the power arguments used to compute the test statistic or p-value. These types of arguments are within arguments. This means that the computations are performed on the same simulated data rather than creating additional simulated data.

Imagine the case of estimating power from competing fitted models. To keep the design relatively simple, two competing models will be explored. One will be the same model that was simulated that includes an interaction effect; the second will omit the interaction when fitting models that were included in the simulated data. Practically, this situation could be of interest for a few different reasons. First, the one attribute that is omitted from the simpler model could be more challenging to collect from individuals. Therefore, there could be more missing data on this attribute, and it could be interesting to explore how much the estimated power could be impacted by having much less data. Secondly, interactions are often more difficult to estimate and have smaller estimated effects. Therefore, the impact on the estimated power could be of interest to determine if the interaction term can be reliably estimated.

9.2.1 Vary Model Fit Arguments

When attempting to vary the model fit arguments, the within arguments are specified the same as the between varying arguments. The **simglm** package aims to simplify this for the user, and auto-detects the model fit (and power arguments, more on this soon) to use these as within arguments. A list of the different model specifications is specified when specifying the model fit arguments. Within each element of the list is one list of model-fitted arguments. Two different model fit arguments need to be specified by default. One is the model formula that is passed to the formula argument of the model fitting function (e.g., `lm()`). The second argument needed is the simulated model's regression weights. Thinking about these in a statistical framework, these would be referred to as the regression weights used to simulate the data. These are needed to specify for the **simglm** package to compute bias for the parameter estimates. Often, these would represent a subset of the regression weights from the simulated generation model. For example, below, the first three are for the terms, intercept, weight, and age for the simpler model and would be specified for the regression weights for the simpler model.

It should also be noted below that the attributes, weight, and age, in this example, are centered to have a mean of zero. This is needed to help increase the interpretation of the intercept but also ensure that the comparison between the reduced and full models. As an interaction between weight and age is used as the population (i.e., within the simulation arguments), then the location of the attributes will impact the parameter estimates and, therefore, power. By centering the attributes at their mean, the interpretations will be comparable across the fitted models for the main effects.

```
set.seed(321)

sim_arguments <- list(
  formula = y ~ 1 + weight + age + weight:age,
  fixed = list(weight = list(var_type = 'continuous',
                             mean = 0, sd = 30),
               age = list(var_type = 'ordinal', levels = -15:15)),
  reg_weights = c(2, 0.3, -0.1, 0.025),
  replications = 5000,
  extract_coefficients = TRUE,
  vary_arguments = list(
    sample_size = c(10, 50, 150),
    error = list(
      list(variance = 15),
      list(variance = 25),
      list(variance = 50)
    ),
    model_fit = list(
      list(formula = y ~ 1 + weight + age + weight:age,
           reg_weights_model = c(2, 0.3, -0.1, 0.025)),
      list(formula = y ~ 1 + weight + age,
           reg_weights_model = c(2, 0.3, -0.1))
    )
  )
)

replicate_simulation(sim_arguments) |>
  compute_statistics(sim_arguments, type_1_error = FALSE)
```

Figure 9.2 shows the estimated power based on 5,000 replications fitting a full and reduced model to the data. Notice differences between the top row (full model) and the bottom row (reduced model) within each term. For example, the power estimates for the full model in the top row are often slightly higher, indicating a higher likelihood of rejecting the null hypothesis for each term. The differences were higher for the intercept and age terms than for the weight term. Therefore, if the research is interested in estimating the age term, there is an increased likelihood of doing this for the full model compared to the reduced model, holding other features constant. There are also higher average

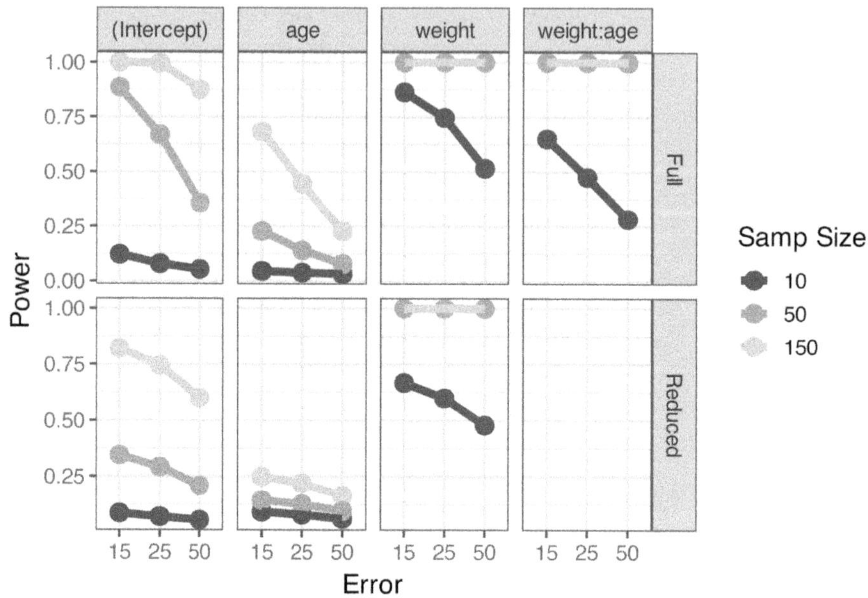

FIGURE 9.2 Estimated power by term, sample size, error variance, and reduced vs full model fitted.

power estimates for the intercept term, although power estimates for the weight term are similar in magnitude across the full and reduced models.

In addition to exploring the power estimates, the simulated data provided avenues for exploring additional outcomes. One outcome that can be interesting to explore is the average parameter estimates for each term. This is computed over all replications within a condition. This outcome can be helpful to compare if the parameter estimates are similar to what was specified in the simulation conditions. If the average parameter estimates differ substantially from the parameters specified in the simulation conditions, this would suggest that the model estimates are likely biased. With biased estimates, the model that is being fitted should be reconsidered.

Figure 9.3 shows the average parameter estimates for the previous example with a full and reduced model fitted to the data. Notice that across the sample size conditions and model terms, there are horizontal bands for each. This reflects that the estimates are consistent regardless of the simulation conditions. Furthermore, these estimates are close to those specified within the simulation arguments, which provides confidence in the power estimate results shown in Figure 9.2.

Another value that could be of interest in this example is the precision ratio in the output from the `compute_statistics()` function. The precision ratio in

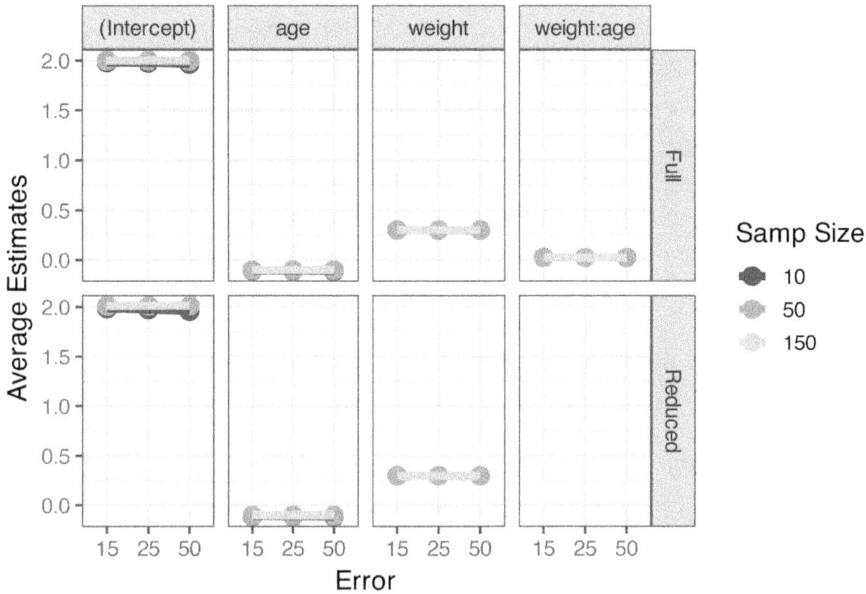

FIGURE 9.3 Average parameter estimates by term, sample sizes, error variance, and full vs reduced model fitted.

this context is defined as the ratio of the standard deviation of the parameter estimates to the standard error estimates. Equation (4.1) shows the computation, where $SD_{estimates}$ is the standard deviation of the estimates and SE is the computed standard error.

$$precision = \frac{SD_{estimates}}{SE} \tag{9.1}$$

The standard deviation of the parameter estimates would represent an empirical estimate for the standard errors computed with the simulated data. The specific interpretation is the average amount of variation that would be expected over repeated simulations for the parameter estimates. Given this ratio, a precision ratio equal to one would indicate that the two estimates for sampling error would be the same. If the precision ratio is greater than one, this would indicate that the standard deviation of the parameter estimates was greater than the computed standard errors. This situation may reflect a situation where the computed standard error is too small, which would, in turn, make the p-value smaller. In terms of power, the power estimate using the p-values may be inflated. In contrast, if the precision ratio is less than one, the standard deviation of the parameter estimates would be smaller than the computed standard error. In terms of power, the power estimate using the p-values may be conservative.

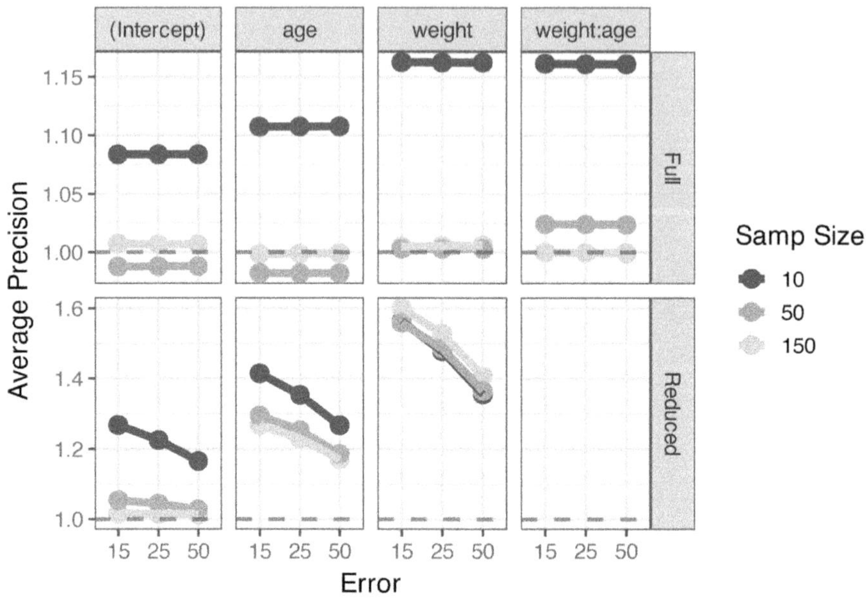

FIGURE 9.4 Average precision ratio by term, sample size, error variance, and full vs reduced model fitted.

Figure 9.4 shows the average precision ratio by the varied simulation conditions. Notice that the precision ratio is close to one for the full model when the sample size is 50 or larger. The precision ratio is greater than one for small sample sizes, which suggests that the variation in the parameter estimates is larger than the computed standard errors. A sample size of 10 is quite small; therefore, this result is not surprising. This result shows that the computed standard errors are similar to empirically computed standard errors when the fitted model matches the data well.

In contrast, the precision ratio statistics are larger for the reduced model and suggest that the computed standard errors are too small overall. If the computed standard errors are too small, it would also produce power estimates using the p-values that would be inflated. Even though the power estimates for the reduced model were smaller than the full model in Figure 9.2, there is evidence to suggest that the power estimates would be inflated. In terms of the magnitude of inflation, the computed standard errors are too small by as much as 60% for the weight term. Using Equation (9.1), if the precision ratio is 1.6 and the average standard error is 0.021, the standard deviation of the parameter estimates could be computed as $1.6 * 0.021 = 0.0336$. The size difference (i.e., average standard error equal to 0.021 vs. the standard deviation of the parameter estimates equal to 0.034) is modest, but this can have an influence on the computed p-values. Methods for using the empirical distribution of the parameter estimates will be presented and discussed in more detail in Chapter 10.

9.2.1.1 Optional Model Fit Arguments

It is also possible to vary the model fitting framework. For example, instead of modeling the mean within the linear regression framework (i.e., `lm()`), it would be possible to model the median using quantile regression. For example, the **quantreg** package and the `rq()` function can model specific quantiles, including the median or 50th percentile. The median is the default quantile to be modeled using the `rq()` function.

The optional `model_function` argument specifies the new model fitting function. These can be specified within any quotations. It can be a good idea to specify the model fitting function using the `packagename::function` syntax to specify explicitly which package the model fitting function should be used for. The package must be installed on the user's computer for the model fitting to work properly. The **simglm** package does not install all model fitting functions by default; these need to be installed by the user.

Any additional optional model-fitting arguments are specified using the default name from the function. For example, within the `quantreg::rq()` function, a model-fitting argument named `tau` specifies the quantile that should be modeled. The default for this function is the 50th percentile, but it is often a good idea to specify this value to be explicit. Notice below that this argument is specified using the default name from the `quantreg::rq()` function. Furthermore, if there are any additional arguments, these are specified within the list as named arguments.

```
model_fit = list(
      list(formula = y ~ 1 + weight + age + weight:age,
           reg_weights_model = c(2, 0.3, -0.1, 0.025)),
      list(formula = y ~ 1 + weight + age + weight:age,
           reg_weights_model = c(2, 0.3, -0.1, 0.025),
           model_function = quantreg::rq,
           tau = 0.5)
    )
```

The updated model code that includes the quantile regression model fitting is shown. For comparison, the prior model fitting process is shown. Notice that the two new arguments, `model_function` and `tau` are specified. As before, the rest of the model fitting is handled by the **simglm** package. If arguments need to be specified in addition to the single optional argument, `tau`, those could be specified within the list as named elements. For example, a `method` argument to the `rq()` function can take on one of ten different estimation methods. The

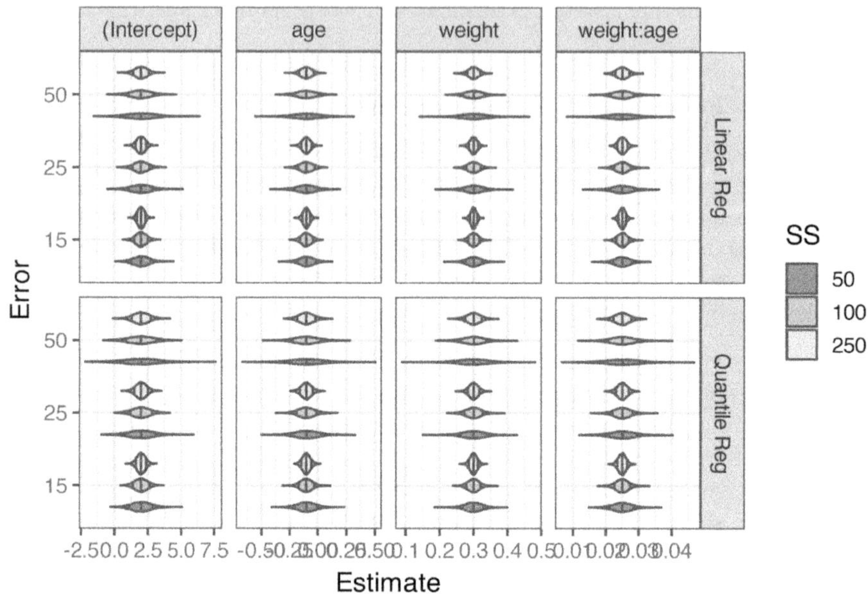

FIGURE 9.5 Distribution of parameter estimates by term, sample size, error variance, and type of model fitted.

default method is "br," but others can be specified and compared for their impact on parameters or power estimates.

The other simulation conditions were kept the same as the previous examples, with one exception. The sample sizes for the simulated data increased compared to the earlier examples. The sample sizes specified were 50, 100, and 250, representing sample size values that would be more realistic.

Figure 9.5 shows the parameter estimates by the error variance, sample size, model fitted, and terms. The 10th, 50th, and 90th percentiles within the violin plots are shown as vertical lines. In this example, the median parameter estimates across the 5,000 replications and all conditions were similar and close to the specified parameters. The primary difference between the two models fitted was in the variance of the parameter estimates. For example, within a panel (e.g., the age term), the violin plot for the parameter estimates is wider for quantile regression compared to linear regression holding other simulation conditions constant.

Table 9.2 shows descriptive statistics for each term and model fit, focusing on a sample size of 50 and an error variance of 25. The mean and median values are similar, but the other statistics show differences. The statistics summarizing variation are larger or more extreme for quantile regression than linear regression. For example, focusing on the term age, the 10th and 90th

TABLE 9.2 Summary statistics for parameter estimates comparing linear regression and quantile regression.

sample_size	error	model_fit2	term	mean	sd	median	min	max
50	25	Linear Reg	(Intercept)	1.997	0.719	1.996	−0.488	5.122
50	25	Linear Reg	age	−0.100	0.081	−0.100	−0.422	0.193
50	25	Linear Reg	weight	0.300	0.025	0.300	0.187	0.416
50	25	Linear Reg	weight:age	0.025	0.003	0.025	0.013	0.036
50	25	Quantile Reg	(Intercept)	1.996	0.905	2.002	−0.952	5.959
50	25	Quantile Reg	age	−0.100	0.102	−0.100	−0.500	0.326
50	25	Quantile Reg	weight	0.300	0.032	0.300	0.151	0.428
50	25	Quantile Reg	weight:age	0.025	0.004	0.025	0.012	0.040

percentiles are about 0.03 larger for quantile regression. The minimum and maximum statistics are also more extreme for quantile regression. Note that the **single** package cannot directly estimate power for quantile regression using the traditional power approach, as no p-values are given. Quantile regression produces confidence intervals, however, which could be used to determine whether the confidence interval contained zero. This is not performed here; instead, it is left to the interested reader to calculate that.

10

Alternatives to Power from Significance Testing

10.1 Chapter Overview

This chapter discusses alternatives to traditional power analysis rooted in null hypothesis significance testing. The chapter introduces simulation-based approaches that focus on estimation accuracy, interval width, bias, and decision-relevant criteria rather than hypothesis rejection. Monte Carlo standard errors and their role in assessing simulation uncertainty are discussed. By broadening the notion of study adequacy beyond statistical significance, the chapter provides tools for researchers whose goals emphasize estimation quality, practical relevance, or policy decision-making. Simulation is positioned as a unifying framework for evaluating these alternative criteria.

Estimating power through simulation methods brings many benefits. One benefit is that the typical power framework from null hypothesis significance testing (NHST) does not need to be followed. When using simulation methods to estimate power, each replication produces an estimate for the population effect of interest. Therefore, the distribution of the estimated effect size (i.e., the sampling distribution of the estimated effect size) can be approximated directly.

The distribution of the estimated effect size can grant many benefits that differ from the NHST framework to discuss elements related to the precision of the effect size. This is often of more interest, particularly when estimating power, as it is often already known that the effect size is not zero. Therefore, rejecting a hypothesis that is already known to be false (i.e., the typical null hypothesis that the population value is equal to zero) is not an interesting statistical or research question. Instead of framing power estimation around a null hypothesis that is different from zero, estimating power by simulation can frame the discussion around precision and how likely the effect size of interest is a practically valuable quantity.

Somewhat more recently, the idea of minimum detectable effect size (MDE) (Bloom, 1995) has gained attention. Although the general idea of the MDE helps frame the discussion around effect size rather than p-values, the idea of the MDE is still primarily focused on the NHST framework. Some recent work has discussed how an MDE is insufficient (Turner et al., 2010); instead,

DOI: 10.1201/9781003177876-10

researchers need to go a step further to articulate that the effect size for the effect of interest is practically relevant to have a substantial impact when implemented.

10.2 Proportion Greater Than a Threshold

When estimating power by simulation, an alternative framework focusing on precision and practically useful effect sizes would be a more robust framework than centering around the NHST. In this scenario, researchers would specify what magnitude of effect size would make a substantial impact. Then, when estimating power by simulation, the alternative power estimate would reflect the proportion of the effect size distribution that would be equal to or greater than the effect size threshold specified (or equal to or smaller than for negative effect sizes).

This statistic of interest here would be the proportion of estimated coefficients that are larger than the specified threshold.

$$Pr(\hat{\beta}_j \geq \tau) \tag{10.1}$$

where $\hat{\beta}_j$ are the estimated coefficients for term j from the simulated data across the number of replications and τ are the thresholds that are defined by the researcher. The statistic represents the likelihood of seeing an estimated effect larger than the practically relevant threshold under the assumed data-generating model. The statistic for negative thresholds would be $Pr(\hat{\beta}_j \leq \tau)$ for a negative threshold τ. Relating this back to traditional NHST power, the proportion of estimated coefficients that are greater than the threshold is similar to that of one-sided power computations.

To summarize this idea in a figure, Figure 10.1 shows the distribution of the estimating effects across the 5,000 replications. Inside the density curve for the distribution of effects, vertical lines are plotted that depict researcher-specific values representing effect sizes that would be deemed practically significant. In Figure 10.1, the terms associated with cloud cover and humidity would be interesting to explore. For these two figures, the alternative power estimate would be the proportion of the 5,000 unique parameter estimates greater than or equal to each vertical line. This proportion can be difficult to estimate from density curves. Fortunately, a helper function within the **simglm** package helps users compute this proportion.

Table 10.1 shows the computed proportion corresponding to the thresholds shown in Figure 10.1. The proportions would be interpreted as the proportion of parameter estimates greater than or equal to each threshold value specified.

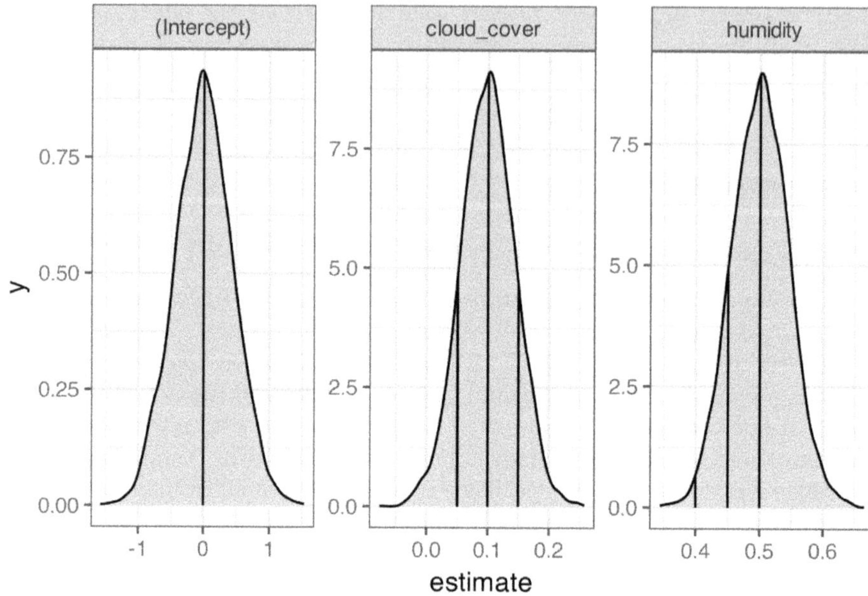

FIGURE 10.1 Alternative power computation based on specified thresholds.

Notice that as the thresholds increase, the proportion decreases, representing fewer overall parameter estimates that would be greater than or equal to that threshold. The proportions would be explicitly tied to the data generated. Therefore, it is important to ensure that the data generation is as accurate as possible for the data to be collected. Often, this means exploring the impact of different data-generating conditions by varying the data-generation conditions.

TABLE 10.1 Alternative power computations for simulated data.

term	avg_estimate	alt_power	threshold	replications
(Intercept)	0.003	0.500	0.00	5000
cloud_cover	0.100	0.875	0.05	5000
cloud_cover	0.100	0.501	0.10	5000
cloud_cover	0.100	0.130	0.15	5000
humidity	0.501	0.990	0.40	5000
humidity	0.501	0.873	0.45	5000
humidity	0.501	0.506	0.50	5000

10.2.1 Identifying Appropriate Thresholds

The challenge in this approach is that the thresholds need to be identified before conducting the research rather than identifying the lower bound of an effect size that would be practically relevant. To develop these values, one would need to have prior experience working in this area or find previous research, perhaps a meta-analysis, to draw on sufficient evidence of standard effect sizes found in the literature. The alternative power framework can be more robust and challenging than the NHST power framework. This is due to the increased argument that needs to be articulated as to why a specific effect size would impact practice. Compare this to the NHST framework, where it is expected to compute if an effect size is statistically different from zero; this can be performed mechanically with almost no scientific thought or relevance.

It is not possible to provide specific thresholds that hold across fields. Instead, the following guidelines aim to give researchers some starting points to establish the argument for why specific thresholds are deemed practically relevant.

1. Look at what effect sizes are in the empirical literature.
   - Pay specific attention to studies that perform well. For example, causal designs would offer stronger evidence.
   - Alternatively, well-done meta-analyses can provide strong evidence for commonly found effect sizes, even without causal designs.
2. Frame the effect size(s) impact on constituents. The following questions help guide that discussion.
   - How would the effect size impact the average constituent?
   - How would the effect size impact other constituents?
   - How would the effect size impact constituents immediately?
   - How would the effect size impact constituents over time?
3. Draft the impact in terms of a policy brief on the program's impact.

10.2.2 Quantile-Based Power Thresholds

It is also possible to take advantage of the distribution of the effect to explore precision. For example, by borrowing the MDE example and translating it to the simulation example, one could compute the percentiles of the distribution of the effect. These percentiles would indicate the threshold for the distribution of the effect that a specific percentage of scores would be less than or equal to that threshold. For example, if the 20th percentile were computed for the distribution of the effect, the value would reflect the estimated effect for which 20% would be less than or equal to that value. In contrast, 80% would equal or exceed this value. This would be similar to the MDE idea, for which a traditional power value of 80% is often specified, identifying the smallest estimate of the effect for which 80% would be equal to or larger than.

TABLE 10.2 Percentiles for the distribution of the estimated effects by the model term.

Outcome	Term	5th Perc	10th Perc	20th Perc
estimate	(Intercept)	−0.733	−0.561	−0.371
estimate	cloud_cover	0.029	0.045	0.063
estimate	humidity	0.427	0.444	0.463

The **simglm** package does not offer a specific function to compute the proportion, although this can be computed directly within R using the `quantile()` function. The `quantile()` function within the `df_stats()` function from the **mosaic** package allows the user to use formula notation and specify the data arguments easily.

```
library(mosaic)

df_stats(estimate ~ term, data = sim_data_df,
         quantile(c(0.05, 0.1, 0.2)))
```

Table 10.2 shows the computed percentiles for the distribution of the estimated effects. The 5th, 10th, and 20th percentiles were computed for each term in the model, including intercept, cloud cover, and humidity terms from the simulated data. In this situation, the intercept is not of direct interest. However, the other two would be the minimum simulated effect size (MSES) associated with a traditional power of 95%, 90%, and 80%, respectively.

To visualize the percentiles, these can be passed as thresholds to plot these on the density curve. Figure 10.2 shows the computed MSES for each term on the distribution of the estimated effects. The thresholds computed would then be compared to theoretical values or in practical terms to understand if the magnitude of the effect would be important practically. More explicitly, the computed thresholds need to be evaluated to determine if they are large enough to have an impact. This is where content knowledge about what is being studied is important, and statistical knowledge is irrelevant here.

10.3 Unconditional Sign Errors

Unconditional sign errors are another useful statistic to consider when estimating power from a simulation based framework. Simulation based frameworks approximate the distribution of the estimated effect size, assuming the data generation process is realistic of the data to be collected. The unconditional

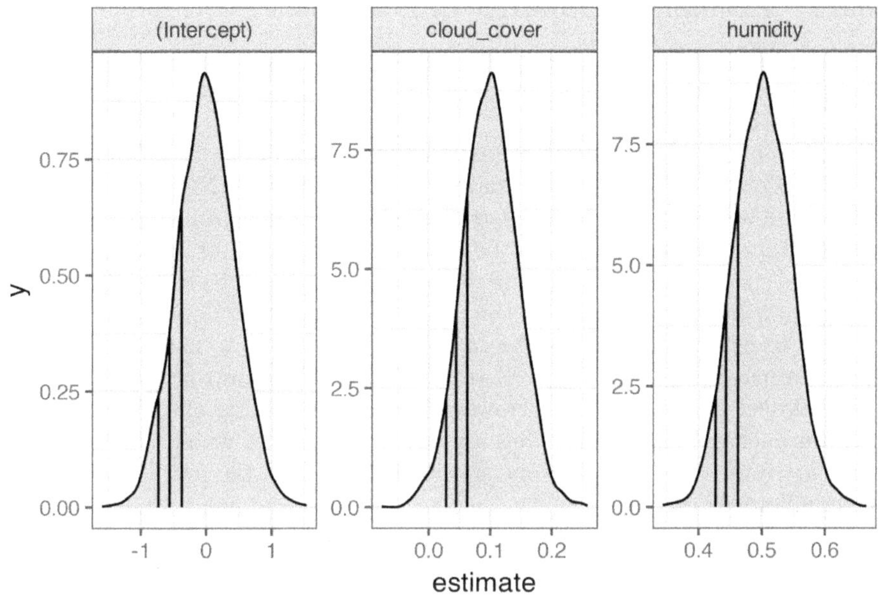

FIGURE 10.2 Minimum simulated effect sizes shown on density figure.

sign errors reflect the proportion of estimated coefficients that are of the opposite sign compared to the population effect size specified. The USE statistic is defined in the following equation:

$$USE = \Pr\big(\,\text{sign}(\hat{\beta}) \neq \text{sign}(\beta)\big) \tag{10.2}$$

where $\hat{\beta}$ are the estimated coefficients across all simulated replications, β represents the population effect size, and *sign* reflects the sign of each associated coefficient. The USE statistic would reflect the likelihood of seeing an estimated coefficient of a different sign than the assumed population effect size. This is estimated by computing the proportion of estimated coefficients have a different sign compared to the specified population effect size. When study planning, this would be an estimate that would want to be minimized as it would represent a case when the study is not able to accurately estimate the direction of the effect size. This statistic is somewhat related to the Type S errors defined by Gelman and Carlin (2014), although the USE statistic is not conditioned on statistical significance.

The **simglm** package offers functionality to compute USE errors for the distribution of the simulated effect and use the `compute_statistics()` function. For USE errors, the direction for the test is specified inside the power arguments with a named list element, `type_use`.

10.3.1 Example Computing USE Errors

To make these errors more concrete, suppose there was a scenario where three groups, low, medium, or high socioeconomic status (SES), were compared after adjusting for a blocking attribute, the individual's age. The outcome here was scores on an anxiety measure where higher scores are indicative of more anxiety. This is an example of analysis of covariance (ANCOVA) discussed in more detail in future chapters. The primary research question focuses on whether there are differences in the three SES groups after adjusting for the individual's age. For simplicity, the data are assumed not to be nested. Finally, the age attribute is generated to be mean-centered. Therefore, a value of 0 is assumed to be the mean age of the sample. This will help to make the intercept the reference SES category. In R, the default SES group is the categorical group closest to the letter A. However, when simulating the data with the **simglm** package, the factor levels are specified directly, which will adjust the factor attribute to be the first one specified. Therefore, the low SES group will be the reference group.

```
##   X.Intercept.     age SES_1 SES_2      SES level1_id
## 1            1  0.2115     0     1 High SES         1
## 2            1  0.3879     0     0  Low SES         2
## 3            1  5.1452     1     0  Med SES         3
## 4            1  1.3827     0     1 High SES         4
## 5            1 -3.7952     0     0  Low SES         5
## 6            1 -2.0606     0     0  Low SES         6
##        error fixed_outcome random_effects anxiety
## 1  1.01254         50.13              0   51.15
## 2 -0.03594         50.16              0   50.12
## 3 -0.66734         51.96              0   51.29
## 4  0.92338         50.60              0   51.53
## 5  1.38110         48.48              0   49.86
## 6  0.87825         49.18              0   50.05
```

The following code expands on the simulation arguments shown earlier by including power arguments. Notice that the thresholds were specified along with the USE statistic signs to be specified. The thresholds for the USE statistic signs are specified by the name of the model term that R uses by default. For continuous data attributes, this would reflect the name specified. However, for categorical or factor data attributes, the default R naming scheme appends the category or factor level to the name. It is advised to run a small simulation example and then extract the names from the term attribute to learn the names to specify. If only some terms are of interest to compute the alternative power or USE statistics, only those terms need to be specified and the others can be omitted. The `compute_statistics()` function can help check whether the correct direction was specified when computing the alternative power statistics;

the sign that was used for computing the USE statistic is shown for users to ensure they are correctly specified.

```
sim_arguments <- list(
  formula = anxiety ~ 1 + age + SES,
  fixed = list(
    age = list(var_type = 'continuous', dist = 'rnorm',
               mean = 0, sd = 3),
    SES = list(var_type = 'factor',
               levels = c('Low SES', 'Med SES', 'High SES'))
  ),
  error = list(variance = 1),
  reg_weights = c(50, 0.4, -0.1, 0.05),
  sample_size = 1000,
  replications = 5000,
  model_fit = list(
    formula = anxiety ~ 1 + age + SES),
  extract_coefficients = TRUE,
  power = list(
    thresholds = list(
      '(Intercept)' = c(0),
      'age'    = c(0.3, 0.4),
      'SESMed SES'  = c(-0.05, -0.1),
      'SESHigh SES' = c(0.05, 0.1, 0.15)
    ),
    type_use_sign = list(
      '(Intercept)' = 'positive',
      'age'    = 'positive',
      'SESMed SES'  = 'negative',
      'SESHigh SES' = 'positive'
    )
  )
)
sim_data <- replicate_simulation(sim_arguments)
```

Notice that the argument, `type_use = TRUE` is specified when using the `compute_statistics()` function in **simglm**. Currently, the USE statistics are the only ones that are being computed, but all of the statistics can be computed or any subset of them.

```
sim_data |>
  compute_statistics(sim_arguments,
                     power = FALSE,
                     alternative_power = FALSE,
```

TABLE 10.3 USE statistic statistics for the simulated example data.

term	avg_estimate	type_use	sign	replications
(Intercept)	50.000	0.000	positive	5000
SESHigh SES	0.049	0.259	positive	5000
SESMed SES	−0.101	0.100	negative	5000
age	0.400	0.000	positive	5000

```
                    type_1_error = FALSE,
                    precision = FALSE,
                    type_use = TRUE)
```

Table 10.3 shows the empirically computed USE statistic statistics from the 5,000 replications of the simulation design. The results show that two effect sizes, the intercept and age effects, did not have any estimated effects flip signs, as shown by the USE statistic rate of 0. However, the two SES effects did have evidence of having effects that flip signs. For the high SES effect, which is the average change in anxiety from the low SES reference group, the effect size is estimated as negative about 26.3% of the time. This would be just over 1,300 of the 5,000 total replications. In the medium SES effect, which is the average change in anxiety from the low SES reference group, the effect size is estimated as positive about 10% of the time. This would be about 500 of the 5,000 total replications. These results are not surprising in relative magnitude as the effect size for the medium SES group was specified as being larger.

Power analysis, including those by simulation, is highly exploratory. What happens if the design is changed so that the age is not collected with the individual, but only the anxiety level and family SES are collected? Prior theory may show that age is an important predictor considering an individual's anxiety level; therefore, this is still specified in the data-generating model that aims to be similar to the real world. The change occurs when specifying the model that is fitted to the resulting data; the variance in anxiety explained by age would be ignored and pooled into the error term. As such, there should be less statistical precision of the SES effects and likely larger USE statistics. Let us explore this impact.

The following code shows the elements of the simulation arguments that would differ from the previous example. Notice that the only pieces that need to change are the `model_fit` simulation arguments and the power arguments to omit the age predictor. The rest of the code as shown before would be the same in this case.

```
sim_args <- list(
  ...,
  model_fit = list(
    formula = anxiety ~ 1 + SES
  ),
  power = list(
    thresholds = list(
      '(Intercept)' = c(0),
      'SESMed SES'  = c(-0.05, -0.1),
      'SESHigh SES' = c(0.05, 0.1, 0.15)
    ),
    type_use_sign = list(
      '(Intercept)' = 'positive',
      'SESMed SES'  = 'negative',
      'SESHigh SES' = 'positive'
    )
  )
)
```

Table 10.4 shows the results from the new model that does not include age in the model fitted to the population data with the age effect included. Notice that the age attribute is no longer included in the table, as this attribute was not modeled. Notice that the average effect sizes for the SES terms are similar to the simulated values. Thus, the data would still estimate the true effect size well, but the rates of a USE statistic increased. About 20% and 32% of the estimated effects for the medium and high SES effects flip signs, respectively. The USE statistic rate for the medium SES term was double what it was with the age attribute included, and the USE statistic rate was 6% larger for the high SES term. It would ultimately be up to the researcher if this would be acceptable, but this highlights that missing key attributes of interest will typically increase the error variance and, thus, the standard errors for terms in the model. Therefore, it will tend to increase the USE statistic rates.

TABLE 10.4 USE statistic statistics for the simulated example data.

term	avg_estimate	type_use	sign	replications
(Intercept)	50.001	0.000	positive	5000
SESHigh SES	0.049	0.342	positive	5000
SESMed SES	−0.101	0.196	negative	5000

10.4 Monte Carlo Standard Errors

Since we are performing Monte Carlo simulation to estimate power, these are simply estimates and asymptotic statistical theory applies. In this situation, more replications, holding everything else constant, will produce a more reliable estimate of power. Similarly, there is a point where increasing the number of replications will not produce a large gain in the reliability of the estimates. Instead, the estimated power will very slowly increase in precision. Monte Carlo standard errors can be a good way to estimate this quantity (Morris et al., 2019). The Monte Carlo standard error is approximated by the following general formula:

$$SE_{mc} \approx \sqrt{\frac{\text{Var}(\hat{\theta})}{R}} \tag{10.3}$$

where R is the number of replications and the numerator is the variance of the statistic of interest. For example, the variance of the proportion would be: $\hat{p}(1 - \hat{p})$. For the USE statistics and alternative power statistics, this can be a good way to approximate how much error there is in the estimate and plan for how many replications to run. For example, if the USE statistic is 0.1 and 2,000 replications are performed, the approximate Monte Carlo standard error would be:

$$SE_{mc} \approx \sqrt{\frac{0.1(1 - 0.1)}{2000}} \approx 0.007 \tag{10.4}$$

This would mean that with 2,000 replications, the approximate Monte Carlo standard error would be about 0.007. If we assume an approximate Normal distribution, we would expect 95% of our statistics to be $\pm 1.96 * SE$ of our statistic of interest, or 0.086 to 0.114. This can then be used to evaluate if this is enough precision for the study planning process or if more replications would be needed. If the number of replications was increased to 5,000, the approximate range would be between 0.092 to 0.108. This is narrower, but may not be worth the extra computational costs associated with them. Computing the approximate Monte Carlo standard error can be a great way to assess level of precision for grant applications or other study planning justification.

If the statistic of interest does not have a well-defined analytic variance, the variance can be approximated directly from the simulated sampling distribution. Specifically, the variance of the statistic computed across Monte Carlo replications can be used as a plug-in estimate in place of a theoretical variance. This approach relies on the fact that, as the number of replications increases, the empirical distribution of the statistic converges to its population sampling distribution. As a result, a larger number of replications may be required to obtain a stable and accurate estimate of the variance of the statistic, particularly for statistics that depend on distributional features such as tails or quantiles. In practice, examining the stability of both the estimated statistic and its Monte Carlo standard error as the number of replications increases can help guide a reasonable choice of R.

11

Randomized Controlled Trials

11.1 Chapter Overview

This chapter applies simulation-based methods to randomized controlled trials, emphasizing the role of randomization in causal inference. The chapter discusses how treatment assignment, compliance, and outcome variability influence power and estimation. Simulation examples illustrate how design features such as sample size, treatment effect heterogeneity, and covariate adjustment affect inferential properties. The chapter demonstrates how simulation can be used to evaluate analytic strategies before data collection, supporting principled design decisions in experimental research.

Randomized controlled trials (RCTs) are often referred to as the gold standard to evaluate treatment effectiveness due to strong internal validity. The strong internal validity stems from randomizing the treatment(s) to those at the intervention level. For example, college students may be randomly assigned to receive tutor support in a remedial math program or receive the business-as-usual condition that allows for office hours with a teaching assistant. The strength of this design is in the ability for the two groups, those receiving the tutor support and those receiving the business-as-usual condition, to be equal, on average, on observed and unobserved data attributes. For example, the math ability, socioeconomic status, or other attributes of the students in this remedial math program would be equal on average.

This has been written about previously, but it should be noted here that the randomization does not guarantee that the groups will be equivalent for *a specific sample.* Put another way, it still is possible for the treatment and control group to be different on observed or unobserved characteristics with random assignment. The randomization operates on expectation, over the long run, or repeated samples. It is often interesting to summarize baseline characteristics to understand chance imbalances and inform precision adjustments (e.g., ANCOVA).

RCTs are one of many designs that should be considered and they can have weaknesses. First, there are many types of interventions or questions for which randomization to treatment is not ethical (e.g., health effects of smoking or effects of bullying). Secondly, RCTs often have poor external validity or

DOI: 10.1201/9781003177876-11

generalization to settings other than those studied. As such, more than one RCT may be needed to evaluate the effectiveness of any treatments in more than one population. RCTs can also be costly to implement. Therefore, it is important to spend time planning sample sizes needed for adequate precision of the treatment effect.

11.2 Common RCT Designs

There are many ways to design an RCT study. It is not possible to explore all possible options for RCT study design. This chapter aims to show how to simulate data from standard RCT designs. The chapter will start simple and then build up to more complicated designs. The three types of RCT designs that will be explored will include:

1. One measurement occasion with treatment effect
2. Pre / Post designs with treatment effect
3. Pre / Post / Follow-up designs with treatment effect

The goal of these types of designs is ultimately to estimate the treatment effect and, in the power framework, to estimate the power of the treatment effect. Furthermore, specific considerations about designing RCTs will not be discussed, such as blinding, other randomization techniques, measuring constructs of interest, or any other study design procedures. One important note is that it is possible to conduct RCTs with more than three measurement occasions. These will be saved for a later chapter, Chapter 13. These will evaluate longitudinal-type designs that can include randomization.

11.2.1 One Measurement Occasion

To build off a simpler example, designing an RCT with a single measurement occasion with randomization to two (or more) groups will be explored. Suppose a pharmaceutical company was developing a new drug to reduce cholesterol in the bloodstream of individuals with high cholesterol. Cholesterol is needed, but too much can increase the risk of heart disease or other ailments. The pharmaceutical company wishes to evaluate the new drug's effectiveness in reducing high cholesterol.

To do this, the company aims to randomize individuals to receive the new drug compared to a placebo condition that does not receive the new drug. As the drug is for individuals with high cholesterol, the population of interest is individuals who have high cholesterol. Suppose, for illustration, the study enrolls participants above a prespecified cutoff of 240 to be high cholesterol and are used as the inclusion criteria for the population of interest. Different inclusion criteria based on the different types of cholesterol could have been made. After taking the new drug (or placebo) for 6 weeks, the company brings the individuals into the lab to test their blood for cholesterol levels.

This design benefits from simplicity; there is only one measurement occasion (i.e., lab visit), which may reduce participant attrition. The design also still implements random assignment for those who take the new drug compared to those who take the placebo. One major weakness of this design is that there is no indication of the level of cholesterol in the individuals prior to receiving the developmental drug. This would limit the ability to evaluate if the random assignment balanced the treatment and control groups regarding the severity of high cholesterol before receiving the developmental drug. As such, it would not be possible to completely rule out that one group may have had lower/higher levels of cholesterol before the randomization.

11.2.1.1 Simulating Data from One Measurement Design

When simulating data from a one-measurement RCT design, the statistical model, in the simplest case, would be.

$$outcome = \beta_0 + \beta_1 treatment + \epsilon \tag{11.1}$$

In this model, the outcome would be whatever outcome is measured. In the example described above about cholesterol, this could be overall cholesterol (or one of the specific types of cholesterol) or whether an individual is below the threshold for high cholesterol (this will be discussed more shortly). The primary term of interest to estimate the average treatment effect would be the β_1 term. The attribute, treatment, would take on a value of one when the individual received the treatment and a value of zero when they received the placebo.

The data would be simulated by specifying the model above. To generate the data with an effect size metric, for example, a Cohen's D or Hedges G statistic, the error variance can be specified as equal to one. This would have the effect of standardizing the error variance. Standardizing the error variance puts effects on a standardized residual scale; with only a binary treatment predictor, the treatment coefficient is interpretable as a standardized mean difference under this simulation framework. When more data attributes are included in the model, those would also need to be included to ensure the metric is standardized. For this example, the effect size is specified at -0.2, meaning that the treatment group has 0.2 standard deviation units less cholesterol than the control placebo condition. The simulation arguments are specified below.

```
sim_arguments <- list(
  formula = cholesterol ~ 1 + treatment,
  fixed = list(
```

```
    treatment = list(var_type = 'factor',
                     levels = c('Control', 'Treatment'))
  ),
  error = list(variance = 1),
  sample_size = 100,
  reg_weights = c(0, -0.2)
)
```

Figure 11.1 shows the distribution of the treatment and control groups for one simulated data. Notice how the treatment condition for the simulated data has a lower median than the control group. This would need to be evaluated with an inferential model to consider sampling error.

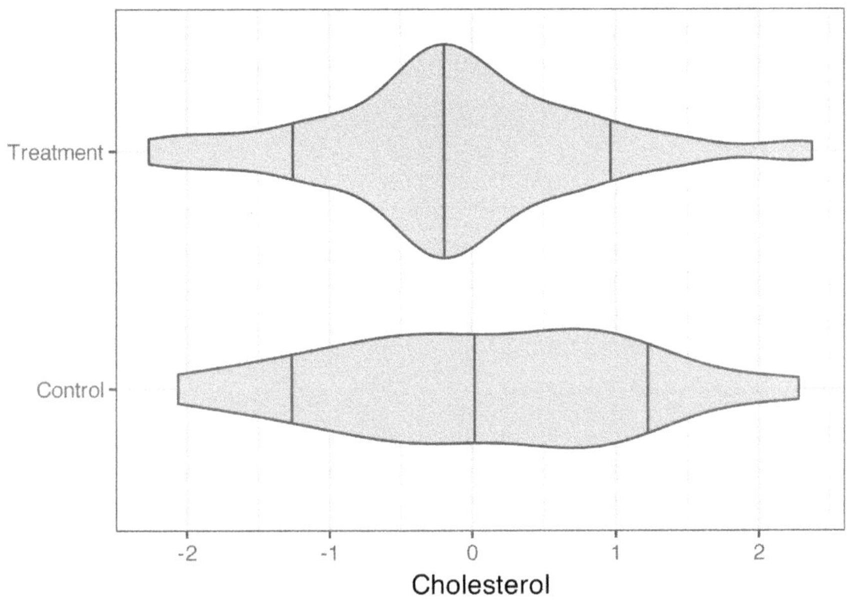

FIGURE 11.1 Distribution of simulated cholesterol post-intervention data by treatment or control groups.

The above design does not consider one element that can be common with treatment conditions: treatment heterogeneity. That is, often, treatment conditions have more variance than the control group after the treatment has been given (Long and Ervin, 2000). Simulating data that have heterogeneity was first shown in Chapter 6. The following heterogeneity simulation arguments can be added to add treatment heterogeneity. These simulation arguments would be simulated within the data using the `simulate_heterogeneity()` function.

Below, the variance of the treatment group is specified to have a variance that is eight times that of the control group.

```
heterogeneity = list(variable = 'treatment',
                     variance = c(1, 8))
```

Figure 11.2 shows the simulated data with the treatment heterogeneity added. The treatment group median is still lower than the control group median; however, the treatment group has many more extreme cases than the control group. When these data are explored with an inferential model, accounting for the heterogeneity (e.g., robust standard errors or weighted least squares) would help ensure standard errors and inferences are robust to the heterogeneity. Ordinary least squares remains unbiased under random assignment, but naive homoskedastic standard errors can be miscalibrated when group variances differ. Alternatively, using the effect size distribution from the simulated data should give a more accurate picture compared to traditional inference with p-values.

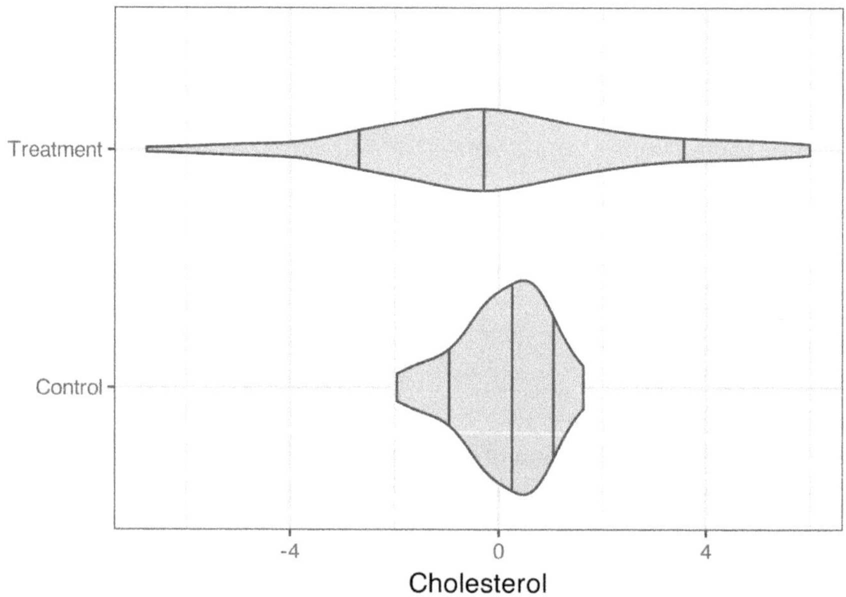

FIGURE 11.2 Distribution of the simulated cholesterol post-intervention data with treatment effect heterogeneity.

11.2.2 Post / Pre Design with Treatment

Adding a pre-test to build a more complicated RCT design than the previous section can increase the design's strength. Adding a pre-test strengthens the design by adding a within-unit component to the design, which can account for where individuals began prior to the treatment condition. For example, using the cholesterol example discussed earlier, adding a pre-test would collect all participants' cholesterol levels before receiving the treatment or placebo.

Even though random assignment is performed with an RCT, this does not ensure that the groups are balanced (i.e., equivalent on all characteristics) for any specific sample. The balance of observed and unobserved attributes described with random assignment is over the long run or in expectation. Therefore, adding the pre-test to an RCT design allows the analyst to adjust for any potential differences in the balance of the sample with any observable characteristics. One observable characteristic that could not have been balanced is the outcome measure before individuals start the treatment. Any unobservable characteristics could still not be statistically adjusted.

The primary weakness of adding a pre-test to an RCT design is the added complexity. For example, researchers need to keep track of records to link the pre-test to the post-test scores. Furthermore, participants must be willing to provide measurements twice instead of once. This can lead to a type of missing data referred to as dropout (Diggle, 2002), where it would be expected that some participants would not return for the second measurement occasion (post-test scores). Perhaps some form of multiple imputation could be successfully used in this situation so as not to need to remove those data records for which the post-test score is missing, but it is best to have the actual measurements rather than using imputation.

11.2.2.1 Simulating Data from Pre / Post Design

A few options exist for estimating the treatment effect from a pre/post-study design. One of these computes difference scores by subtracting the pre-test score from the post-test score, $post - pre$. The second method is to use analysis of covariance (ANCOVA). When analyzing pre/post-study designs with ANCOVA, the post-test score is the outcome, and the pre-test score is used as a covariate to adjust for where participants started. The ANCOVA method will be unbiased when random assignment works well and will also work if the groups are not balanced on the pre-test scores. Both options are unbiased under randomization; ANCOVA is typically more efficient when baseline and post are correlated (Senn, 2006).

The ANCOVA model in the simplest design is as follows.

$$Chol_{post} = \beta_0 + \beta_1 Chol_{pre} + \beta_2 treatment + \epsilon \tag{11.2}$$

In Equation (11.2), $Chol_{post}$ is the measured cholesterol after the treatment, $Chol_{pre}$ is the measured cholesterol before the treatment, and *treatment* is the attribute that tracks whether an individual received the experimental cholesterol drug or not. The parameter of interest would be β_2, representing the average treatment effect.

```
sim_arguments <- list(
    formula = chol_post ~ chol_pre + treatment,
    fixed = list(
        chol_pre = list(var_type = 'continuous',
        mean = 0, sd = 1),
        treatment = list(var_type = 'factor',
        levels = c('Control', 'Treatment'))
    ),
    error = list(variance = 1),
    sample_size = 100,
    reg_weights = c(0, 0.5, -0.25)
)
```

Figure 11.3 shows the simulated data from the RCT simulated through an ANCOVA model. The figure shows that the median post-test score for the

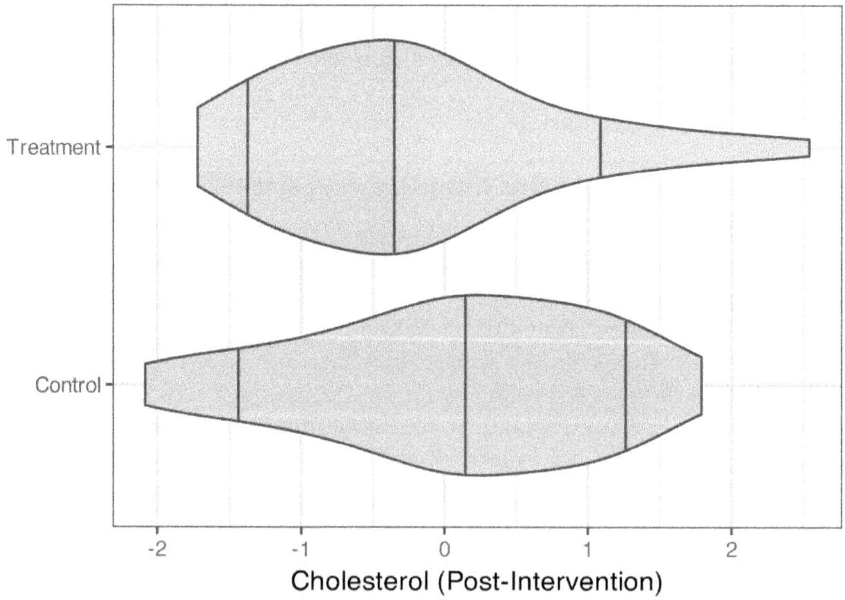

FIGURE 11.3 Distribution of simulated cholesterol post-intervention data by treatment or control group.

treatment is less than the control group. The average pre-test scores for the treatment and control group are 0.21 and 0.21, respectively. The default simulation behavior is to have the pre-test score independent (i.e., uncorrelated) with the treatment attribute. Practically, this means that the default simulation method is for random assignment to simulate balanced data over the long run (i.e., the treatment and control group have the same pre-test score, on average).

It is possible to generate data where the pre-test score is correlated with the treatment group. To do this, the attributes specified as predictors need to be correlated with one another when simulated. The `correlate_variables()` function helps with this process. To use this function, a "correlate" simulation argument is specified, which specifies which attributes should be correlated and their strength. The direction of the correlation specified controls which group will have higher scores on the pre-test measure. A positive correlation will mean the group dummy coded as one will have higher pre-test scores and vice versa for a negative correlation.

The "correlate" simulation arguments are shown below as a named list. As the correlated attributes are fixed effects, a data frame specifies the pair of attributes to be correlated along with the correlation coefficient. The idea of correlated attributes was discussed more in Chapter 6.

```
correlate = list(fixed = data.frame(x = c('chol_pre'),
                                    y = c('treatment'),
                                    corr = c(0.6)))
```

The correlation was positive so that the treatment group would have higher pre-test scores. For example, with a correlation of 0.6, the average pre-test cholesterol scores for the treatment and control group are 0.47 and −0.31, respectively. As the two groups differ on the pre-test scores, this would represent a case when the random assignment was not performed or when the random assignment did not work well for the sample collected.

11.2.3 Post / Pre / Follow-up with Treatment

Another addition that can be common with RCT study designs is the addition of a follow-up time after the post-test. The timing of when the follow-up would occur would depend on what is being measured, but the overall goal or question would be related to how well the treatment would persist beyond the post-test measurement occasion. Often, in these study designs, exploring trends within a longitudinal framework is not of interest; rather, the follow-up measurement occasion is compared to either the pre-test or post-test scores to evaluate if the impact of the treatment declines over time. Evaluating long-term effects is often expected when the intervention is something that is learned. For example, a newly designed behavioral intervention to reduce symptoms of obsessive-compulsive disorder (OCD) may be explored via an RCT. The intervention sessions may stop after many occasions for the new behavioral intervention. The follow-up measurement occasion can evaluate if the skills learned via the behavioral intervention continue to reduce symptoms of OCD six months after the treatment stopped.

When evaluating follow-up measurement occasions within an RCT, the research question would guide the analysis that is done. For example, if the question is whether the treatment loses effectiveness compared to immediately after the treatment, then the follow-up measurement occasion would want to be compared to the post-test score. Instead, if the treatment offers lasting benefits, then the comparison may want to be done about the pre-test score.

When considering an RCT study with a follow-up measurement occasion, significant planning to limit participant attrition for the follow-up measurement occasion should be considered. This can be a severe limitation of the follow-up measurement occasion, especially if the follow-up measurement occasion occurs long after the post-test measurement occasion. Along with attrition, the type of attrition that occurs should be considered. For example, are those who still need to complete the follow-up measurement occasion less enthusiastic about the treatment or had higher (or lower) pre- or post-test scores? These situations indicate that the attrition is not random but depends on other participant characteristics. This needs to be appropriately modeled in the analytic stage.

11.2.3.1 Simulating Data from Pre / Post / Follow-up Design

Multiple ways would be possible to simulate data from an RCT design that wishes to collect data via follow-up. Ultimately, the simulation strategy would need to mirror the research questions; therefore, the focus on what comparison to make (pre-test or post-test measures) must be clear before designing the simulation. Once this is specified, much of the simulation process is the same as previous examples in this chapter.

Suppose the research question for the follow-up measurement occasion is to compare if the intervention improved OCD symptoms compared to baseline (i.e., pre-test measure). The follow-up was designed to be measured six months after the behavioral intervention ended. In this situation, the comparison or covariate to be included would be the pre-test for both of the model comparisons. The following would represent the two models that could generate the simulated data.

$$symp_{post} = \beta_0 + \beta_1 symp_{pre} + \beta_2 treatment + \epsilon \tag{11.3}$$

$$symp_{followup} = \beta_0 + \beta_1 symp_{pre} + \beta_2 treatment + \epsilon \tag{11.4}$$

These two equations would represent the two different comparisons made, one for the post-test and another for the follow-up measurement occasions. In both, the OCD symptomatology at the pre-test measurement occasion and the treatment parameter (β_2) would represent the average treatment effect for post-test and follow-up, respectively.

To simulate the two equations simultaneously, the formula and regression weights simulation arguments are specified as a list where each element represents the formula and regression weights for each equation, respectively. The **simglm** package verifies if the fixed attributes are the same across each formula only to generate the fixed elements once. Then, the specific regression weights, in tandem with the fixed portion of the simulation formula, are used to generate each outcome. The following code chunk shows how these two new elements are specified. The treatment effect for the follow-up was specified to be different from the post-test effect for illustrative purposes. Often, the follow-up effect would be smaller than the post-test effect.

```
formula = list(
      symp_post ~ symp_pre + treatment,
      symp_followup ~ symp_pre + treatment
  )
reg_weights = list(
      c(0, 0.5, -0.25),
      c(0, 0.5, -0.10)
  )
```

Table 11.1 shows the first few rows of the simulated RCT data with a follow-up effect. Since this was simulated in one process, the same individuals received the treatment for the post-test and follow-up measurement occasions. This would mirror what would be done in practice. If the simulation was conducted in two stages, one step simulating the post-test and a second step simulating the follow-up, the treatment effects would be accurate, but the data would not directly reflect what is collected in practice. For power estimation, either approach should yield similar results. That is, if the goal is to estimate power

TABLE 11.1 First few rows of RCT data with a post-test and follow-up simulated.

symp_pre	treatment	symp_post	symp_followup
−0.512	Treatment	0.03961	0.190
−0.144	Treatment	1.13221	1.282
−1.050	Control	0.51220	0.512
0.960	Control	0.69411	0.694
1.177	Treatment	1.00459	1.155
−2.012	Treatment	−1.18762	−1.038

TABLE 11.2 Descriptive statistics for simulated RCT data.

Outcome	Treatment	Mean	SD	Median	IQR	N
symp_post	Control	0.415	0.797	0.518	1.046	45
symp_post	Treatment	−0.090	0.957	−0.212	1.150	55
symp_followup	Control	0.415	0.797	0.518	1.046	45
symp_followup	Treatment	0.060	0.957	−0.062	1.150	55

for post-test and follow-up measurement occasions within an RCT, the power estimates should be similar if the power estimates are generated from simulated data in a single step compared to simulated separately.

Finally, Table 11.2 calculates descriptive statistics for the simulated follow-up data. The table shows the descriptive mean differences for the post-test and the follow-up measurement occasions based on the treatment allocation. Two things to notice from this table. First, the number that were randomly assigned the treatment are the same for the follow-up and post-test. Secondly, the control group has the same descriptive statistics. Together, these two points show that the same individuals are assigned the treatment and control conditions for the post-test and follow-up measurement occasions. The average treatment effects are (roughly) shown as the difference between the control and treatment groups for each outcome. This is why the treatment effect was specified to be larger in magnitude at post-test (more negative) and smaller at follow-up.

11.3 Estimate Power for RCTs

A study by Blom et al. (2015) will be used to evaluate the power to plan for a future RCT trial on the levels of depression for caregivers of those with dementia. The study by Blom et al. (2015) evaluated the effectiveness of a course called Mastery over Dementia to improve depression or anxiety symptoms for caregivers of those with dementia. The study implemented a pre/post design to evaluate the course's effectiveness in reducing caregivers' symptoms. The course consisted of eight lessons and a booster session a month after the eight primary lessons for those randomized to receive the treatment condition. As part of the course, each individual received a coach who offered feedback to the caregivers on homework assigned as part of the lessons. The control group received newsletters by email every three weeks for six months.

Overall, Blom et al. (2015) reported that the course reduced depression and anxiety symptoms by 0.29 and 0.34 standard deviation units, respectively. Tables 3 and 4 of the original study show the unstandardized and standardized treatment effect estimates. Using this information, one could design the simulation using standardized or unstandardized metrics. The two should give similar information, but the primary benefit of the unstandardized metric is that it directly interprets the metric of the depression or anxiety measurement tool used. The standardized metric may be helpful if a different measure is used for depression or anxiety. For example, Blom et al. (2015) used the Center for Epidemiologic Studies Depression Scale and the Hospital Anxiety and Depression Scale, anxiety subscale, to measure depression and anxiety, respectively, but the researchers of the current study are aiming to use a new researcher made instrument. In the latter case, thinking about the effect size magnitude from other studies in the standardized metric would be more relevant.

11.3.1 Standardized Metric

When simulating data in the standardized metric, the value for the effect size would be entered for the effect size. In this case, the effect size sign would be flipped from what was reported in the original study, as the treatment effectiveness would reflect a *reduction* in depression symptoms for caregivers. Procedurally, this means that the average post-test depression symptom score would be lower than the average pre-test depression symptom score. This was entered below as -0.29 as the final regression weight.

The code below would generate data for one simulation with a sample size of 100. Testing the simulation design with a single iteration is a helpful first step before jumping to many different replications. Exploring this simulated data descriptively can offer an important step to ensure that the simulated data roughly matches the population characteristics. With a single replication, these will not match the population characteristics perfectly, but the characteristics should be close to the population values.

```
sim_arguments <- list(
    formula = dep_post ~ dep_pre + treatment,
    fixed = list(
        dep_pre = list(var_type = 'continuous',
        mean = 0, sd = 1),
        treatment = list(var_type = 'ordinal',
        levels = c(0:1))
    ),
    error = list(variance = 1),
    sample_size = 100,
    reg_weights = c(0, -0.5, -0.29)
)

dep_data <- simulate_fixed(data = NULL, sim_arguments) |>
  simulate_error(sim_arguments) |>
  generate_response(sim_arguments)
```

TABLE 11.3 Summary statistics for simulated pre/post RCT data.

Outcome	Treatment	Mean	SD	Median	10th perc	90th perc	N
dep_post	0	0.078	1.363	−0.182	−1.474	2.116	39
dep_post	1	−0.256	1.093	−0.146	−1.747	1.077	61
dep_pre	0	−0.319	1.114	−0.225	−1.706	0.991	39
dep_pre	1	−0.189	0.860	−0.087	−1.339	0.838	61

Table 11.3 shows the descriptive statistics of the pre and post outcomes by the treatment status. Overall, there is an average decrease for the treatment group for the post test scores, which would reflect the specified direction of the effect size. This would not match the exact specified effect size, but at a minimum should likely reflect the same direction. The descriptive mean difference may not be in the same direction and the simulation may still be specified appropriately, this would then be a case that would reflect a USE statistic or if conditional rather than unconditional means are more appropriate.

11.3.1.1 Designing Simulation Conditions

The next step in estimating power for the RCT example would be to define specific simulation conditions that may impact the power estimates. This could be varying the effect size, sample size, balanced vs unbalanced treatment assignment, or adding treatment effect heterogeneity. Given that random assignment is used in an RCT, the treatment and control groups would want to be equivalent. Therefore, the design would not want to specify a correlation between treatment and the pre-test scores. When these are simulated as not being correlated or independent, then on average, the treatment and control groups will be equivalent at baseline.

Suppose the following simulation conditions were of interest to explore: 1. effect size magnitude: −0.15, −0.20, −0.25, −0.30, −0.35, −0.40 2. sample size: 50, 100, 150, 200, 250 3. treatment allocation rates: 30%, 40%, 50% treatment allocation 4. effect size heterogeneity: 3, 5, 8 times larger variance for treatment group.

Commonly, all of these conditions would be crossed or a factorial design. The total number of conditions being evaluated would be multiplicative. Therefore, there would be $6 * 5 * 3 * 3 = 270$ total simulation conditions. This needs to be considered as it will take time to simulate the data, fit the model, and extract statistics for all conditions and replications. For example, if 5,000 replications were specified (a strong starting point for stable Monte Carlo estimates), this would result in $270 * 5000 = 1,350,000$ total simulated data sets, fitted models, and extracted statistics.

If time was a factor, the number of simulation conditions or replications could be reduced. The main concern for each would be less coverage of data collection procedures that may occur in practice or, with fewer replications, less precision estimating power. These may be valid trade offs, but should be considered when designing the study to estimate power.

11.3.1.2 Estimating Power

The replicated statistics from the simulated data and the fitted models can be used to estimate power. First, it can often be informative to descriptively explore the distribution of the estimated effect sizes. To simplify the discussion, only a single population effect size was chosen for initial analysis. The effect size of -0.30 was chosen and the distribution of the estimated effect sizes are shown in Figure 11.4. The estimated effect size was consistently estimated with each distribution centered around -0.30. Furthermore, there is increased precision in estimation as sample size increases. Treatment effect heterogeneity and a smaller proportion receiving the treatment increases the amount of variation in the distribution of the estimated effect size. Although not depicted, the distribution of the estimated effect size was similar across the different population effect sizes.

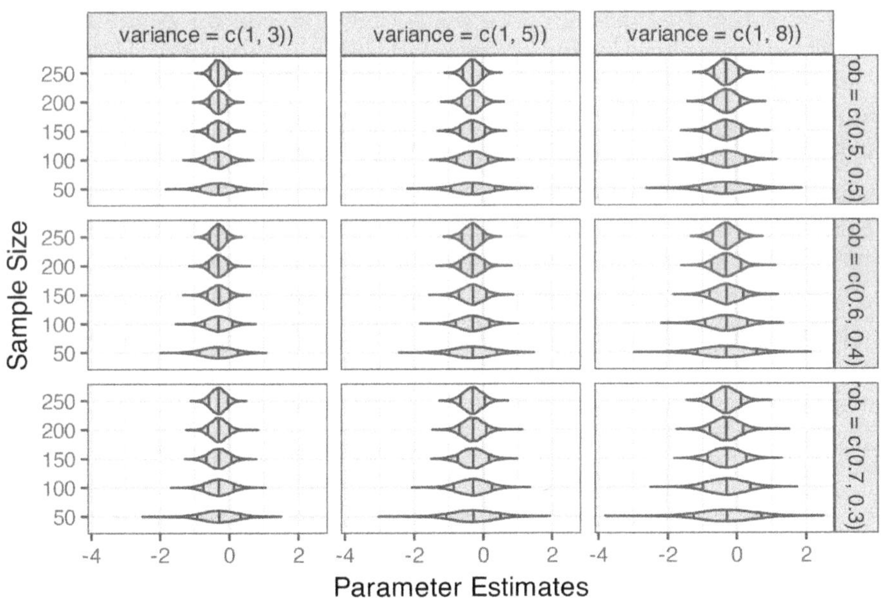

FIGURE 11.4 Distribution of the estimated effect sizes across 5,000 replications by sample size, treatment heterogeneity, and proportion that received the treatment.

The focus of estimating power will be on minimizing any USE statistics and evaluating the proportion of estimated effect sizes that are greater than specified thresholds. This framework utilizes some of the benefits of the simulation approach to estimate power and also can help minimize the statistical data conditions that are needed for valid statistical inference (i.e., normally distributed residuals, heterogeneity of variance, etc.) The traditional power analytic framework can also be used. This will be shown briefly, but interpretation will be less of a focus. Some comparisons comparing the estimates between the two approaches will be made.

Estimating USE Statistics

When designing a study, USE statistics would represent cases where the sign would flip. USE statistics would want to be minimized, especially for studies that are confirmatory, when study planning as it would represent an effect that is not accurately estimating whether the treatment is positive or negative. The purpose of the study would need to be considered to determine acceptable levels of USE statistics for the research study.

Table 11.4 shows the ten largest estimated USE statistics. Not surprising, the estimated USE statistics were large for the smallest sample sizes and effect size conditions. The effect of the treatment heterogeneity and proportion of respondents in the treatment condition are shown. For example, when the variance of the treatment group is eight times the control group, the USE statistic rates are about 5% higher compared when the variance of the treatment group is three times larger than the control group. Furthermore, the effect of proportion receiving the treatment is about 3% smaller when the groups are equal compared to only 30% receiving the treatment.

Figure 11.5 visually depicts the USE statistics by sample size, population effect size specified, and the amount of treatment effect heterogeneity (shown in the figure panels) when treatment and control groups are equal in size (i.e., 50%

TABLE 11.4 USE statistic statistics for the treatment effect of the simulated example data. The data are arranged to show the ten largest USE statistic rates.

SS	Beta	Fixed	Het.	Term	Avg. Est	USE	Sign	Repl
50	−0.15	0.7, 0.3	1, 8	Treatment	−0.137	0.429	negative	5000
50	−0.15	0.6, 0.4	1, 8	Treatment	−0.149	0.410	negative	5000
50	−0.15	0.7, 0.3	1, 5	Treatment	−0.139	0.407	negative	5000
50	−0.2	0.7, 0.3	1, 8	Treatment	−0.187	0.402	negative	5000
50	−0.15	0.5, 0.5	1, 8	Treatment	−0.153	0.397	negative	5000
50	−0.15	0.6, 0.4	1, 5	Treatment	−0.149	0.393	negative	5000
100	−0.15	0.7, 0.3	1, 8	Treatment	−0.150	0.389	negative	5000
50	−0.15	0.7, 0.3	1, 3	Treatment	−0.141	0.385	negative	5000
50	−0.2	0.6, 0.4	1, 8	Treatment	−0.199	0.383	negative	5000
50	−0.2	0.7, 0.3	1, 5	Treatment	−0.189	0.380	negative	5000

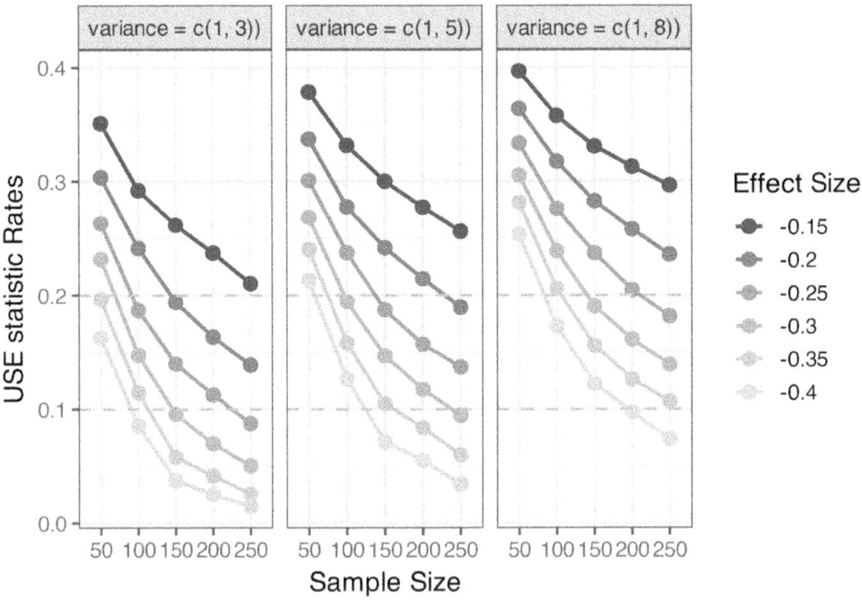

FIGURE 11.5 USE statistics for the treatment effect by treatment heterogeneity, population effect size, and sample size. The treatment and control groups were specified to be equal.

in treatment and control). The figure helps to show how the estimated USE statistics for larger effect sizes decrease faster when increasing sample sizes. The effect of treatment heterogeneity is also quite large and should not be minimized when study planning. When doing traditional closed form equations to estimate power, homogeneity of variance is assumed. This will be explored even more in future sections, but the increase in USE statistics shows an increase in variation for the treatment effect, which will reduce the estimated power.

The horizontal lines in Figure 11.5 are set at 0.1 and 0.2, representing 10% and 20% USE statistic rates, respectively. These can be adjusted to reflect the USE statistic rates that could be accommodated when planning the research study. The 10% rate was shown here as it would represent a smaller USE statistic rate for confirmatory type research and 20% may be a good value for exploratory work to adequately identify the direction of the estimated effect.

Estimating MSES

When estimating MSES, the thresholds that are deemed practically relevant need to be specified. For the treatment effect, these were specified as -0.1, -0.2, -0.3 as three practically useful and relevant quantities. For instance,

TABLE 11.5 Minimum simulated effect size (MSES) for the treatment effect of the simulated example data. The data are arranged to show the ten largest MSES rates.

SS	Beta	Fixed	Het.	Term	Avg. Est	MSES	Threshold	Repl
250	−0.4	0.5, 0.5	1, 3	Treatment	−0.398	0.949	−0.1	5000
250	−0.4	0.6, 0.4	1, 3	Treatment	−0.406	0.944	−0.1	5000
200	−0.4	0.5, 0.5	1, 3	Treatment	−0.397	0.930	−0.1	5000
250	−0.4	0.7, 0.3	1, 3	Treatment	−0.406	0.925	−0.1	5000
200	−0.4	0.6, 0.4	1, 3	Treatment	−0.403	0.921	−0.1	5000
250	−0.35	0.5, 0.5	1, 3	Treatment	−0.348	0.912	−0.1	5000
250	−0.4	0.5, 0.5	1, 5	Treatment	−0.398	0.906	−0.1	5000
150	−0.4	0.5, 0.5	1, 3	Treatment	−0.405	0.904	−0.1	5000
250	−0.35	0.6, 0.4	1, 3	Treatment	−0.356	0.904	−0.1	5000
250	−0.4	0.6, 0.4	1, 5	Treatment	−0.408	0.896	−0.1	5000

the thresholds may represent small, medium, or large effect sizes. These would need to be developed with the specific problem context in mind.

Table 11.5 shows the ten largest MSES statistics for the treatment effect. The largest values MSES statistics are for the smallest threshold, largest population effect size specified, the largest sample sizes, and with the smallest amount of treatment heterogeneity. In these simulation conditions, over 90% of the estimated effects were smaller than −0.1 (i.e., larger effects).

Figure 11.6 shows the MSES statistics visually. This figure shows the MSES statistics by the sample size, population effect size, treatment effect heterogeneity in the top panels, and the thresholds in the side panels. Horizontal lines are shown at estimated power values of 0.65 and 0.80. These two values would represent cases that the estimated effect is larger (i.e., more negative in this example) in at least 65% and 80%, respectively. If the population effect size is larger than 0.25 or so, all conditions have estimated MSES values that are greater than 0.65 for the smallest threshold. For effect sizes that are greater than −0.3 and sample sizes greater than 200, the estimated MSES values are above 0.80.

As the thresholds increase, the MSES statistics tend to be smaller than the −0.1 threshold. This means that these situations occur less frequently, but there are still instances when the estimated MSES are greater than 0.80 and 0.65. These do occur with larger population effect sizes and less treatment effect heterogeneity. This is important context to consider when planning a study to ensure that the desired effect size can be estimated with some level of frequency. Given that the smallest threshold is one that is deemed to be practically relevant, this would be the one to give the most weight to when

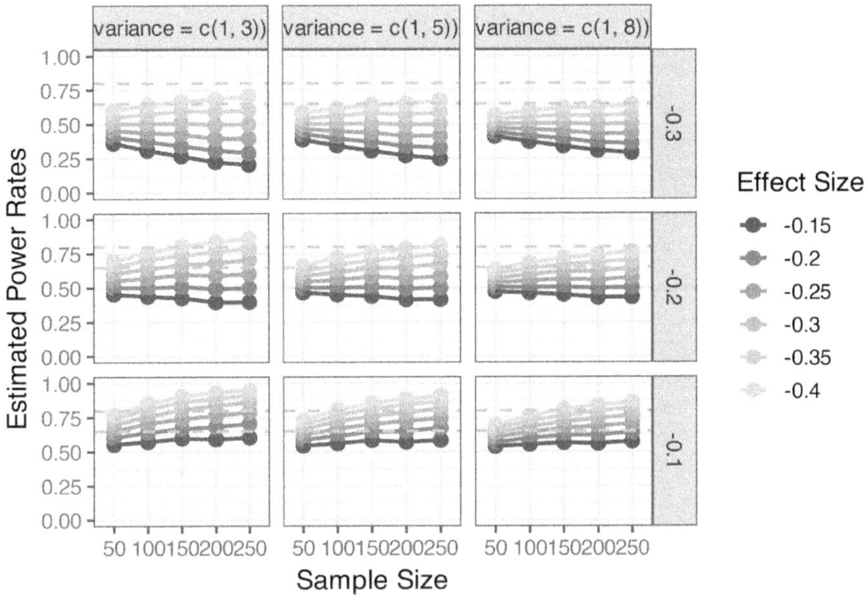

FIGURE 11.6 MSES statistics by the sample size, population effect size, treatment effect heterogeneity (top panels), and thresholds (side figure panels). The MSES statistics are shown for equal probability of being in the treatment group.

study planning. Although, it is helpful to see the rates of larger effect size estimates as a way to gauge potential impact.

One might wonder why some of the MSES statistics decrease as sample sizes increase. This occurs for the population effect sizes specified for the largest thresholds. Ultimately, the reason this occurs is due to the increased level of precision in estimating the treatment effect with larger sample sizes. Given that the smallest population effect size was specified at -0.15 and the threshold for a large effect size was -0.30, the increased precision makes an effect size as large as -0.30 more difficult to achieve. The increased precision is a good thing to accurately estimate the treatment effect, however, the increased precision makes it more unlikely to detect the large effect threshold if the true effect size is smaller.

Combining USE Statistic and MSES Information

To be fully informed, it is useful to combine the USE statistic and MSES information. If possible, minimizing risk of detecting USE statistics and maximizing the MSES values would be a good way to successfully plan a study. There could be trade offs needed depending on budget and other implementation considerations, however. For example, if the budget would only allow for so

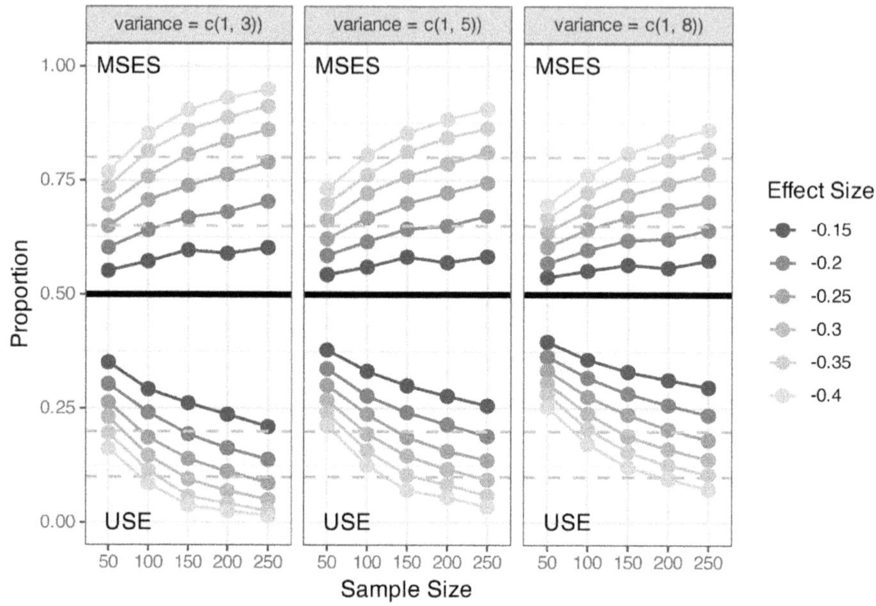

FIGURE 11.7 Combining Type S and MSES statistics into a single figure, USE statistics on the bottom half and the MSES statistics on the top half. The threshold for MSES shown is for the smallest minimum effect of practical utility (−0.1).

many to be assigned to the treatment condition, that may impact the ability to increase sample size drastically. Time for study implementation may be another limiting factor when selecting sample sizes too.

For the specific example simulated, the effect of treatment heterogeneity needs to be considered when integrating information from the two statistics. USE statistics were not a large concern when treatment heterogeneity was small, but were more problematic with large amounts of treatment heterogeneity. Figure 11.7 combines the two statistics into a single figure to compare. This figure shows the MSES above the solid horizontal line and the USE statistics below the solid horizontal line. In this figure, the best conditions would be those that diverge the most to be highest on the top half and smallest on the bottom half. For example, the best conditions across all of the treatment effect heterogeneity are those that have the largest population effect size and larger sample sizes. In contrast, the conditions with the smallest population effect size show much less divergence indicating that the USE statistics and MSES statistics are similar. These statistics would reflect those that estimating the correct direction of the treatment effect occurs less frequently and estimating an effect size at least as large as −0.1 happens slightly over 50% of the time.

TABLE 11.6 Traditional power estimates for the treatment effect of the simulated example data. The data are arranged to show the ten largest estimated power rates.

SS	Beta	Fixed	Het.	Term	Avg. Est	Power	Repl
250	−0.4	0.6, 0.4	1, 3	Treatment	−0.406	0.532	5000
250	−0.4	0.7, 0.3	1, 3	Treatment	−0.406	0.521	5000
250	−0.4	0.5, 0.5	1, 3	Treatment	−0.398	0.489	5000
200	−0.4	0.6, 0.4	1, 3	Treatment	−0.403	0.433	5000
200	−0.4	0.7, 0.3	1, 3	Treatment	−0.399	0.433	5000
250	−0.35	0.7, 0.3	1, 3	Treatment	−0.356	0.432	5000
250	−0.35	0.6, 0.4	1, 3	Treatment	−0.356	0.426	5000
250	−0.4	0.7, 0.3	1, 5	Treatment	−0.407	0.422	5000
250	−0.4	0.6, 0.4	1, 5	Treatment	−0.408	0.400	5000
200	−0.4	0.5, 0.5	1, 3	Treatment	−0.397	0.388	5000

Estimating Traditional Power

It is also possible to estimate the proportion of times the null hypothesis of zero was rejected. This is the traditional null hypothesis significance testing framework. Table 11.6 shows the ten largest estimated power values. In this instance, the increased precision of larger sample sizes for the largest population effect sizes has the largest estimated power values. Even though these are the largest, the treatment heterogeneity may have lowered the estimated power, but the typical gold standard of 80% was not achieved.

Figure 11.8 shows the estimated power values from the NHST framework using power curves. The largest increases are driven by sample size and treatment effect heterogeneity. For example, these power curves show the impact treatment effect heterogeneity can have on rejecting the null effect of zero, with estimated power staying close to or less than 0.25 when treatment effect heterogeneity is eight to one. Sample sizes would need to be increased if this was deemed a possible situation that is concerned about. If treatment effect heterogeneity is less of a concern, then larger sample size than 250 would be desired to reject the null effect when the population effect sizes are above -0.3.

11.4 Using a Meta-Analysis to Estimate Power

Meta-analyses are another way to estimate the effect size. Meta-analyses combine statistical information from more than one study to provide a better estimate of the effect size. Meta-analyses can also evaluate if there is treatment

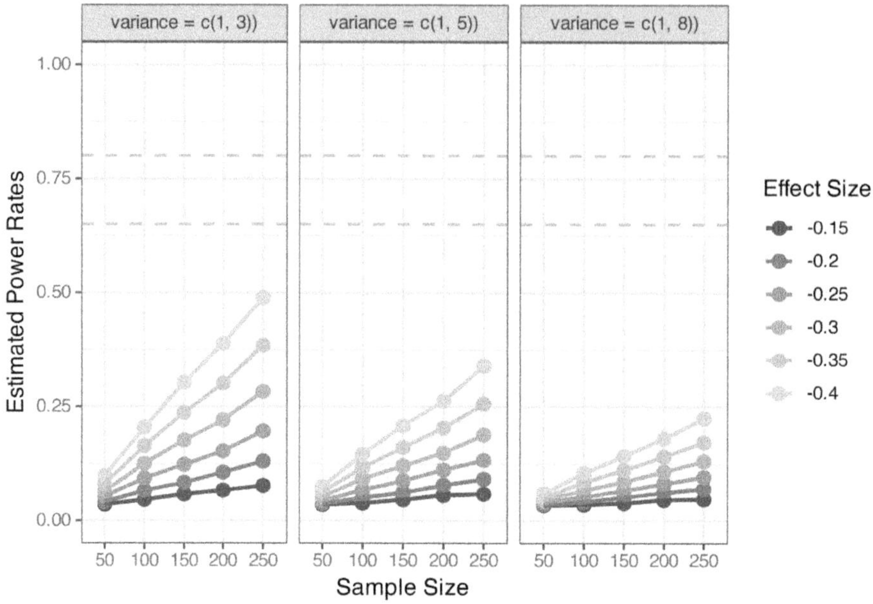

FIGURE 11.8 Estimated NHST power by the sample size, population effect size, treatment effect heterogeneity (top panels), and thresholds (side figure panels). The MSES statistics are shown for equal probability of being in the treatment group.

heterogeneity or variation in the treatment effect across studies. For example, is there evidence that the treatment effect is similar across different studies, or is there evidence that the treatment effect size differs across studies? This is important information to consider when designing a simulation to estimate power. If there is evidence of treatment effect size heterogeneity, it would be important to vary the treatment effect size estimate in the simulation design to account for this heterogeneity. In contrast, if there is evidence of more homogeneous treatment effect sizes, the effect size used in the simulation design would not need to vary as much. There would be exceptions to this; for example, if the study that is being planned is being done on a different population from the meta-analysis, then it would be best to still vary the treatment effect size to account for potential population differences.

In the context of the example related to depression of caregivers of those with dementia, a meta-analysis was performed by Etxeberria et al. (2021). This meta-analysis included the study by Blom et al. (2015) when estimating the aggregated effect size estimate. Figure 4 of the Etxeberria et al. (2021) shows the combined studies to come up with the average effect size estimate of -0.213 (note, the meta-analysis used Hedges's g effect size rather than Cohen's D of the original study). This estimate still shows that the average effect size favors the treatment group, which shows that the treatment is effective in reducing depressive symptoms in caregivers of those with dementia.

When exploring Figure 4 in the Etxeberria et al. (2021) study, three of the five studies favored the treatment group, but two had empirical estimates that favored the control group. Furthermore, four of the five studies had large effect size standard errors with an extensive range of plausible values for the treatment effect. The one exception to this was the study by Blom et al. (2015), which had larger sample sizes and a more precise estimate of the treatment effect. This was given more weight when estimating the aggregate effect size in the meta-analysis.

Given the range of plausible values from the meta-analysis, ranging from −0.03 up to −0.40, it would be worth simulating the treatment effect size within this range. If the assumption is that the treatment is effective, then the minimum effect size would not be specified down to −0.03. Instead, picking a value deemed practically useful would be a good effect size to specify as the lower bound. For hypothetical purposes, a value of −0.10 will be used as the smallest effect size to offer practical utility. Effect size values that range down to −0.40 by intervals of 0.05 (i.e., −0.10, −0.15, −0.20, and so on) will be specified.

Much of the mechanics of simulating the data would follow from the example simulated using estimates from Blom et al. (2015). However, the simulation would need to be utilized on the standardized metric due to how the average effect size was computed as part of the meta-analysis. The remaining mechanics would look very similar.

12

Cluster Randomized Controlled Trials

12.1 Chapter Overview

This chapter focuses on cluster randomized controlled trials, where intact groups rather than individuals are assigned to treatment conditions. The chapter discusses the implications of clustering for power, estimation, and interpretation, emphasizing the role of intraclass correlation and cluster size. Simulation examples illustrate how design choices such as the number of clusters and within-cluster sample sizes affect statistical efficiency. The chapter highlights how simulation provides a practical approach for evaluating complex clustered designs that are poorly served by simple analytic formulas.

Cluster randomized controlled trials (C-RCTs) are similar to RCTs discussed in Chapter 11 in that units are randomly assigned to treatment(s) or control groups. However, the primary difference is the unit that is being randomized. In C-RCTs, the randomization unit is an intact cluster or group instead of individuals. For example, within education, it is often not feasible to randomly assign students within a classroom to receive a different curriculum. Instead, the randomization is done at the school or classroom level, and all students within the school or classroom would receive the treatment condition. Often, the desired unit of analysis within C-RCTs is the unit within the clusters. Therefore, the data clustering needs to be considered in the analysis.

C-RCTs can still have strong internal validity due to the random assignment, but since the randomization is being performed at the cluster level, often more data is needed to achieve the same level of control and statistical precision, and therefore comparable power, as in RCTs. For C-RCTs, with few clusters, chance imbalance is more likely and treatment effects are estimated less precisely, so power can be substantially lower than in individually randomized designs. In the simulation framework used here, this shows up as lower probabilities of exceeding practically useful thresholds (MSES) and higher probabilities of wrong-sign estimates (USE). Much of the same drawbacks are found within C-RCTs compared to RCTs, namely high cost, potentially poor external validity or generalizability, and not all questions are ethically able to be studied using a C-RCT design. Further information about study planning for the C-RCT can

DOI: 10.1201/9781003177876-12

be found at the following resources (Hayes and Moulton, 2017; Raudenbush, 1997).

12.2 Simulating Data from C-RCTs

Simulating data from C-RCTs is similar in style to RCTs in that often C-RCTs involve one measurement occasion, two measurement occasions (often pre- and post-intervention), or could contain three measurement occasions (pre-, post, and follow-up). Using C-RCTs with more than three measurement occasions would be possible, but that discussion will be saved for Chapter 13.

The critical addition to simulating data from C-RCT designs is through nesting and random effects. Instead of having a single level of analysis, as was the case when RCTs were discussed in Chapter 11, a second (or third) level of analysis is added where the randomization occurs. For example, a professional development training program for teachers was established to increase knowledge of advanced coursework options for students. In this situation, it may be possible to randomize directly with teachers and evaluate their knowledge. One drawback of this approach could be teachers sharing the knowledge gained via the training with other teachers who needed the training in the same school. This could be a threat to the fidelity of the training program. As such, it may be desirable to randomize at school instead of teachers to lessen the impact of inadvertently sharing the training knowledge.

The randomization unit would be schools, and all teachers within that school would receive the training. The primary outcome could be teacher knowledge of advanced coursework options for students. Therefore, the unit where inferences are hoped to be made differs from the level where the randomization occurs. As such, it would be essential to account for the nested nature of the data, either modeling the school level with a random effect or including fixed effects for schools. Furthermore, the example uses a pre- and post-intervention measurement strategy. The knowledge of advanced coursework options for students is evaluated prior to the professional development and then measured after the professional development program.

The ANCOVA mixed model that could match this design is as follows.

$$knowledge_{post} = \beta_0 + \beta_1 knowledge_{pre} + \beta_2 PD + b_{0s} + \epsilon \tag{12.1}$$

In this equation, the knowledge score post-intervention is the outcome of interest, and the knowledge pre-intervention adjusts for the amount of knowledge that teachers started with. The primary attribute, PD, represents whether a school was given professional development, and the associated term, β_2, would help estimate the average treatment effect. When simulating the data, this term would be simulated at the school level instead of the teacher level. Adding the parameter, b_{0s}, represents the school random effect, where the s subscript indexes schools. Conceptually, this term represents school-specific deviations or whether a school would be below or above the average school knowledge of advanced coursework options.

Recall from Chapter 3 that random effects are specified using model syntax similar to the **lme4** package. To add the school-level random intercept, this would be specified as `knowledge_post ~ knowledge_pre + PD + (1 | school)`. The other simulation argument that would be added would be specifying what the random effect distribution should be simulated from. Recall that this is done with the named simulation argument `randomeffect`. This simulation argument is specified in the code chunk below. The variance of the random effect is specified as equal to one, and the level this is simulated at is level 2 (teachers nested in schools).

Furthermore, the sample size must contain levels one (i.e., teachers) and two (i.e., schools). In the code below, the number of schools was specified as 30, and the number of teachers within each school was set to vary from 10 to 50 for each school. This would mimic schools that are of varying sizes.

```
randomeffect = list(
  int_school = list(variance = 1, var_level = 2)
)
sample_size = list(level1 = sample(10:50, size = 30, replace = TRUE),
                   level2 = 30)
```

Figure 12.1 shows the simulated data for post-intervention knowledge for those schools that received the professional development (PD) compared to schools that did not receive the professional development. The simulated effect size was 0.45, suggesting that receiving the PD resulted in post-intervention knowledge that was 0.45 units larger than those that did not receive the PD. Figure 12.2 is an alternative visualization that shows the PD effects and variation across schools. On average, the schools that received the PD have teachers with increased knowledge compared to the control group, but there is variation in both conditions. When analyzing these data, it is important to consider and account for clustering. If ignored, they will be absorbed into the residual variance and reduce the precision of the estimated treatment effect. This will be discussed in more detail in the estimating power section of this chapter.

12.2.1 Randomization at a Third Level

The previous example had the randomization occur at the second level, but it is possible to have randomization occur at another level, for example, the third level. Suppose there is a city-wide intervention program that is aimed at reducing violence. The outcome of interest is the number of violent acts

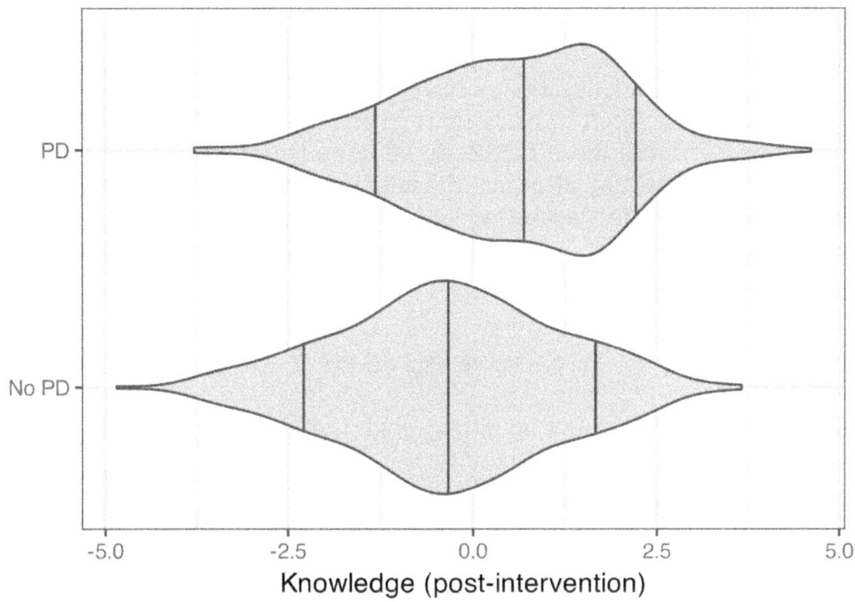

FIGURE 12.1 Distribution of simulated knowledge post-intervention data by professional development group from C-RCT design.

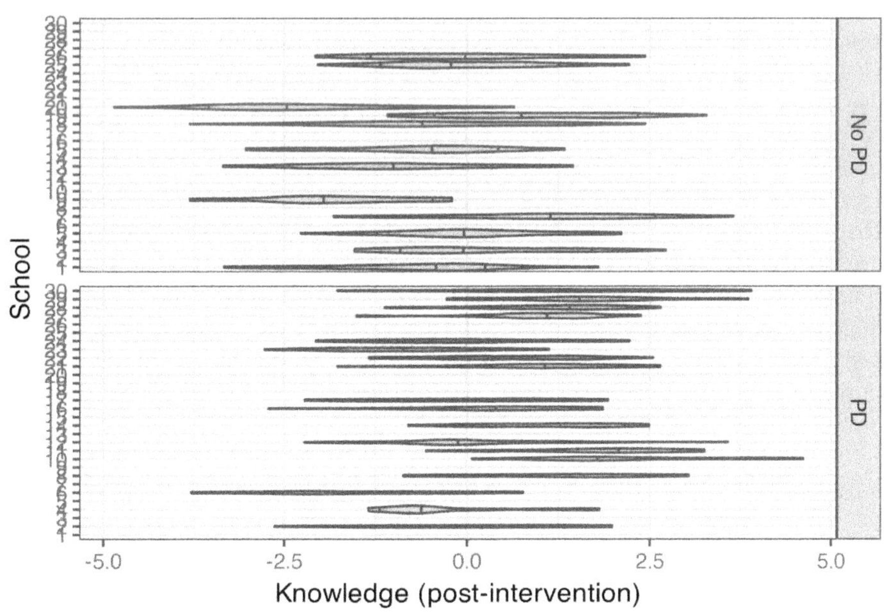

FIGURE 12.2 Distribution of simulated knowledge post-intervention data by schools randomized to professional development group from C-RCT design.

six months after the city-wide intervention. The design also adjusted for the number of violent acts six months before the city-wide intervention.

In this case, the randomization occurred at the city level. That is, entire cities were randomly selected to participate in the intervention to reduce violence. Cities can be broken down further to be more hyper-local. The study design separated city units and identified the number of violent acts into the following levels: neighborhoods, regions, and the city level.

The model for this type of situation could build off the mixed ANCOVA model specified in Equation (12.1)

$$violence_{post} = \beta_0 + \beta_1 violence_{pre} + \beta_2 CI + b_{0rc} + b_{0c} + \epsilon \tag{12.2}$$

Many terms are the same as above, but there are a few differences. The CI term is whether or not the city was randomized to be included in the intervention. Therefore, this attribute would be simulated at the city level (i.e., level three). There are now two random effects, one for the city level (b_{0c}, with c indexing cities) and another at level two (b_{0rc}, with r indexing regions within the city). The two violence measures, one pre-intervention, and the outcome post-intervention would be measured at the neighborhood level. The β_2 term would represent the average city treatment effect.

The model equation to simulate this three-level mixed model is: `violence_post ~ violence_pre + CI + (1 | region) + (1 | city)`. There are now two random effects specified that would represent the b_{0rc} and b_{0c} from Equation (12.2). This specification assumes region IDs are unique across cities; otherwise use a nested term (e.g., city/region). Furthermore, since the outcome is the number of violent acts, the outcome type is specified as counts with the simulation argument `outcome_type = 'count'`. Traditionally, there is no ϵ, or error term, when using a count outcome, but in the simulation framework, **simglm** allows the user to add additional unexplained error variance that is added to the linear predictor (on the link scale). After this is added, the transformation function is applied to turn the outcome into the count metric, which can have the effect of adding overdispersion to the outcome. Similarly to specify the number of violent acts in the previous six months as an integer (i.e., count metric), the "violence_pre" attribute was specified with the following characteristics:

```
violence_pre = list(var_type = 'continuous',
                    dist = 'rpois', lambda = 115)
```

The random effects were very similar to the previous example. The level three city random effect was added, which was specified with a variance of 0.5, and attribute level three. Finally, the sample sizes were specified where there were ten cities; within each city, there could be between 3 and 8 regions, and within each region, there are between 3 and 15 neighborhoods. It should be noted that these values were not taken from any real cities; therefore, these may be unrealistic. Regardless, the sample size arguments would use the following code.

```
set.seed(10)

level2_ss <- sample(3:8, size = 10, replace = TRUE)
level1_ss <- runif(sum(level2_ss), min = 3, max = 15) |> round(0)

sample_size = list(level1 = level1_ss,
                   level2 = level2_ss,
                   level3 = 10)
```

Figure 12.3 shows the number of violent acts for those that were in the city-wide intervention (labeled as "CI" in the figure) compared to the cities that did not receive the city-wide intervention. The figure shows fewer violent acts in the six months after the intervention for those cities that used the intervention. Figure 12.4 shows violin plots for each city based on whether they were in the city-wide intervention. The figure shows variation across cities, but on average,

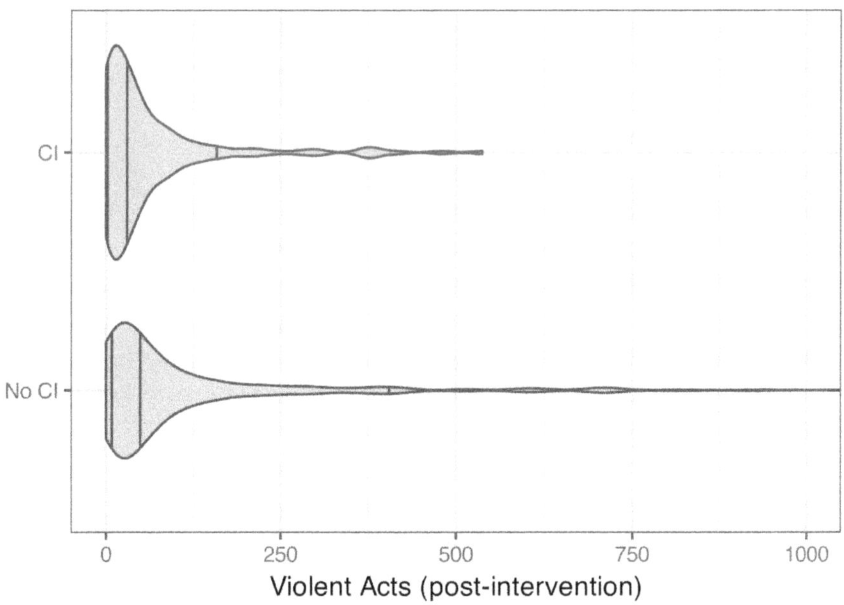

FIGURE 12.3 Distribution of simulated violent acts post-intervention data by city-wide intervention groups from C-RCT design.

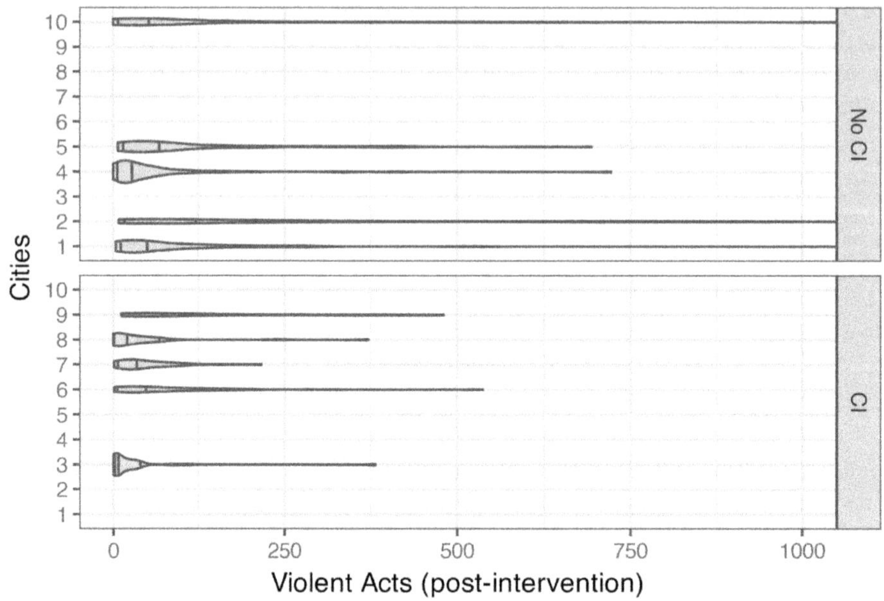

FIGURE 12.4 Distribution of simulated violent acts post-intervention data by cities randomized to city-wide intervention groups from C-RCT design.

the city-wide intervention did have a descriptive effect on reducing the number of violent acts over the six months after the intervention.

12.3 Evaluating Power for C-RCTs

A study by Neuman et al. (2011) will be used to show how power can be estimated for a C-RCT. The Neuman et al. (2011) study implemented a vocabulary intervention aimed at improving preschool word knowledge and concept development. As is common in education, it was likely not possible to randomize students as the student already belonged to an intact group, classrooms. It would be incredibly challenging for a classroom teacher to alter their teaching for individual students. Furthermore, it would not be possible to eliminate crossover effects where the treatment could also influence students who were randomized to the control condition. For these reasons randomization either occurs at the classroom or school to mitigate crossover effects or increase treatment implementation.

The study by Neuman et al. (2011) randomized the vocabulary intervention, named World of Words, to schools. All students and classrooms that participated in the study within each school received the treatment or control conditions. That is, there were no cases where the treatment and control conditions were both implemented within a school. The analytic design followed the pre- and post-test design across different vocabulary units, units one through three. The power example that follows will focus on their unit one and word knowledge assessment. Power estimation would follow similarly for the other units and outcomes.

12.3.1 Simulate C-RCT Data

Table 9 from the Neuman et al. (2011) study contains the information from the statistical modeling and can help us to create simulated C-RCT data that would mirror the real world. The treatment effect for word knowledge for unit one was estimated to be 0.29, indicating that those in the treatment group have better word knowledge following the treatment compared to the control group. This treatment effect was standardized, therefore, the 0.29 increase is in terms of standard deviation units in the outcome. This can make planning the simulation easier. All the fixed attributes in the simulation design would be simulated to have a standard deviation of one and the error variance would also be equal to one. The other fixed effect that was modeled in the original study was an expressive vocabulary measure, the Woodcock-Johnson picture vocabulary measure. The pre-test vocabulary measure was shown to be positively related to the post-test word knowledge measure, but the treatment did not moderate this association.

For the C-RCT data, the study design was nested and a random effect would need to be added to simulate this type of design. The study randomized the treatment condition to schools; however, the study used classrooms within the schools as the nested attribute for students. This does represent a mismatch between the randomization unit (i.e., schools) and the attribute that was modeled (i.e., classrooms). For the word knowledge outcome, a random intercept was specified (see Table 9 from Neuman et al. (2011)). The standard deviation of the random classroom intercept was estimated at 0.22 (variance would be 0.05).

To keep the example two-level and align the treatment level with the modeled clustering, we treat the classroom as the randomized cluster (e.g., selecting one classroom per school or assuming schools contribute one effective cluster). A fuller representation would add a third level (school) and assign treatment at the school level.

These data were specified with the following selected simulation arguments. One key element that is different from the Chapter 11 is the specification of the treatment effect. Notice that the optional argument, `var_level = 2` is specified. This indicates that the treatment effect is associated with the classroom and that all the level one units (i.e., students inside the classroom) would all receive the treatment condition. This is the appropriate unit for a C-RCT where the randomization occurs at the cluster level.

```
formula = knowledge ~ 1 + wj_pre + treatment + (1 | classroom)
fixed = list(
    wj_pre = list(var_type = 'continuous', var_level = 1,
                  mean = 0, sd = 1),
    treatment = (list(var_type = 'factor', var_level = 2,
                 levels = c('Control', 'Treatment')))
  )
randomeffect = list(
    int_classroom = list(variance = 0.05, var_level = 2)
  )
```

Figure 12.5 shows the distribution of the word knowledge scores for the treatment and control group. Notice the slight shift to the right for the treatment group compared to the control group, which would represent the positive treatment effect. That is, there are higher word knowledge scores on average, compared to the control group. This is a descriptive effect that also does not consider the pre-test scores, but would represent an example of a

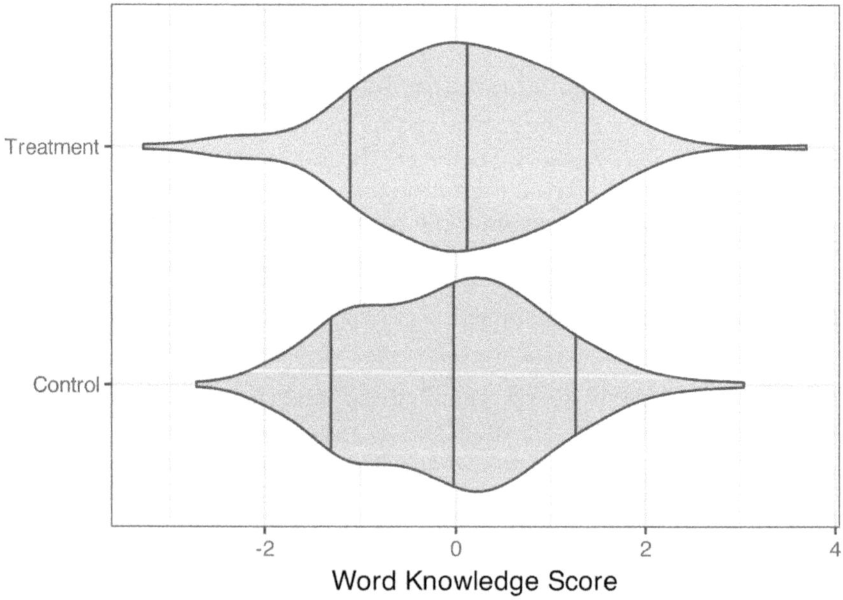

FIGURE 12.5 Violin plots of word knowledge scores for the treatment and control groups. 10th, 50th, and 90th percentiles are shown as vertical lines.

figure to explore before the modeling phase to estimate the average treatment effect. When simulating data, this figure can provide verification that the simulation arguments were specified correctly regarding the average treatment effect.

12.3.2 Estimate Power for C-RCT Data

When using the study by Neuman et al. (2011) to estimate power for future study planning, the average treatment effect and variance of the classrooms can be good baseline values. Estimating the variance of the clustering attribute can be challenging to estimate and varying this is a good way to understand the impact it can have on the estimated power. The intraclass correlation (ICC) can be used to calculate the percentage of variance that is at the cluster level. The ICC represents a ratio of variances and reflects the variance attributable to the cluster level divided by the total variance. For a two-level model, the general form for the ICC is: $ICC = \tau^2/(\tau^2+\sigma^2)$ where τ^2 is the variance for the cluster and σ^2 is the within cluster variance. For example, if the variance for level one was equal to 1 and the cluster level variance was equal to 0.05 (as specified in the simulation example), then the ICC would equal $0.05/(1 + 0.05) = 0.048$. This would indicate that about 4.8% of the total variance is at the classroom level and 95.2% of the variance is at the student level. Given that the 0.05 is an estimate from the sample data from Neuman et al. (2011), this would be an important condition to vary when estimating power for study planning.

Other elements to consider varying when study planning for a C-RCT would be, sample sizes (both level one and two), effect size strength, inferential model fitted, effect size heterogeneity, simulation generating model, or additional levels in the study design. To start, the following simulation arguments will be considered to vary:

- Effect Size Magnitude: 0.20, 0.25, 0.30, 0.35, 0.40
- Classroom Variance: 0.075, 0.10, 0.15
- Number of Classrooms (level 2 sample size): 20, 30, 40, 50
- Analysis Model Fitted: linear mixed model vs linear model with cluster robust standard errors.

The number of classrooms was varied, but not the level one (number of students in the classroom) as in this type of study design, the number of classrooms could be directly added/removed. However, it would be more difficult for the researcher to change the number of students in the classroom. There could be other instances where the level one sample size could be impacted, for example, if a study was implementing workshops as a treatment condition, the number in the workshop could be changed within the study planning.

The varying arguments would be specified as the following when designing the simulation to estimate power.

```
vary_arguments = list(
    sample_size = list(list(level1 = level1_ss_20, level2 = 20),
                       list(level1 = level1_ss_30, level2 = 30),
                       list(level1 = level1_ss_40, level2 = 40),
                       list(level1 = level1_ss_50, level2 = 50)),
    reg_weights = list(c(0, 0.15, 0.20),
                       c(0, 0.15, 0.25),
                       c(0, 0.15, 0.30),
                       c(0, 0.15, 0.35),
                       c(0, 0.15, 0.40)
    ),
    randomeffect = list(
      list(
        int_classroom = list(variance = 0.075, var_level = 2)
      ),
      list(
        int_classroom = list(variance = 0.1, var_level = 2)
      ),
      list(
        int_classroom = list(variance = 0.15, var_level = 2)
      )
    ),
    model_fit = list(
      list(formula = knowledge ~ 1 + wj_pre + treatment +
             (1 | classroom),
           model_function = lme4::lmer,
           name = 'lme4'),
      list(
        formula = knowledge ~ 1 + wj_pre + treatment,
        model_function = lm,
        name = 'robust',
        robust = TRUE,
        robust_args = list(cluster = 'classroom')
      )
    )
  )
```

12.3.2.1 Distribution of the Treatment Effect

The first step in evaluating power is to again ensure that the average estimates are unbiased and representative of the population being simulated in this case. Figure 12.6 shows that the estimated statistics are on average equal to the specified population values. This shows that the expectation worked well in this case and that there are no issues with the simulation process.

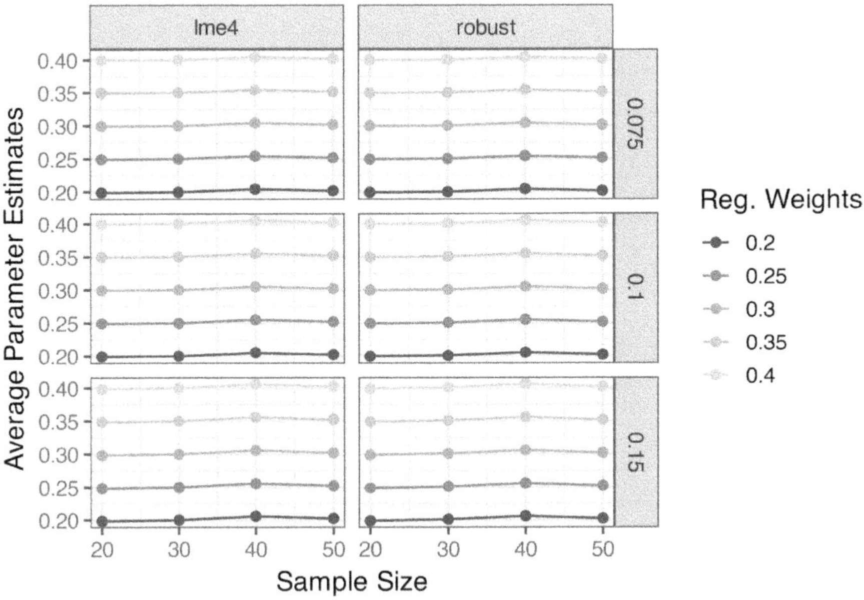

FIGURE 12.6 Average simulated model estimates by level two sample size, term, regression weights, random effect variance, and model fitted.

Figure 12.7 shows the estimated MSES statistics by all of the simulation conditions that were varied for the mixed effect specified models. The robust standard error approach yielded very similar results and therefore are not shown visually, but subsequent interpretations would not change for those models.

The power estimates do not change drastically across the three top panels showing the amount of random effect variance. The estimated power is slightly lower as the random effect variance increases, but this is not large enough to drastically change the interpretation across the different random effect variance combinations explored. There is more impact with regard to the level two sample size and the magnitude of the population effect size specified (right panels). With a few exceptions, the estimated power does not go above 0.65 for those when the population effect size is 0.25 or smaller. The estimated power approached 0.80 when the population effect size was specified as 0.3 and the smallest threshold value, but the estimated power is smaller for larger thresholds. For the largest two specified population effect sizes, the estimated power is within typically defined ranges for power, at or above 0.80. This occurs for most study conditions and for most researcher defined thresholds.

The USE statistics shown in Figure 12.8 are generally modest in nature across most study conditions. The one noticeable exception to this is when the

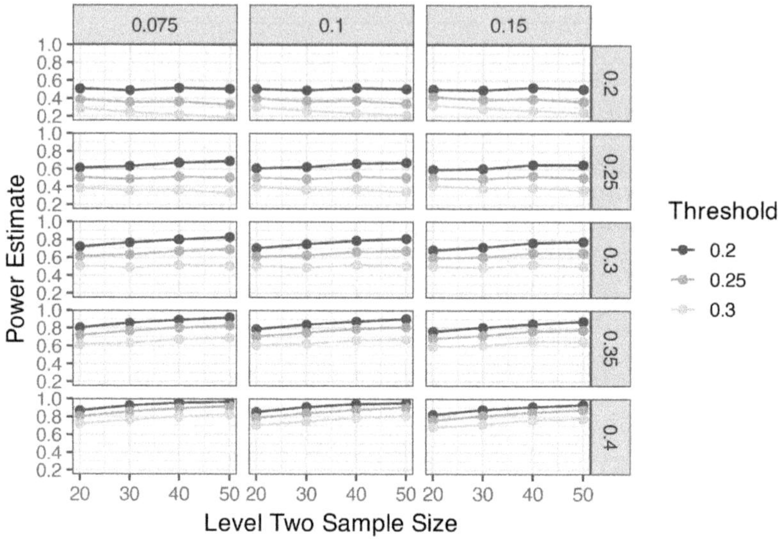

FIGURE 12.7 MSES statistics using the three specified thresholds for practically useful effect sizes by level two sample size, regression weights (right panels), and the random effect variance (top panels) for the mixed effect specified model.

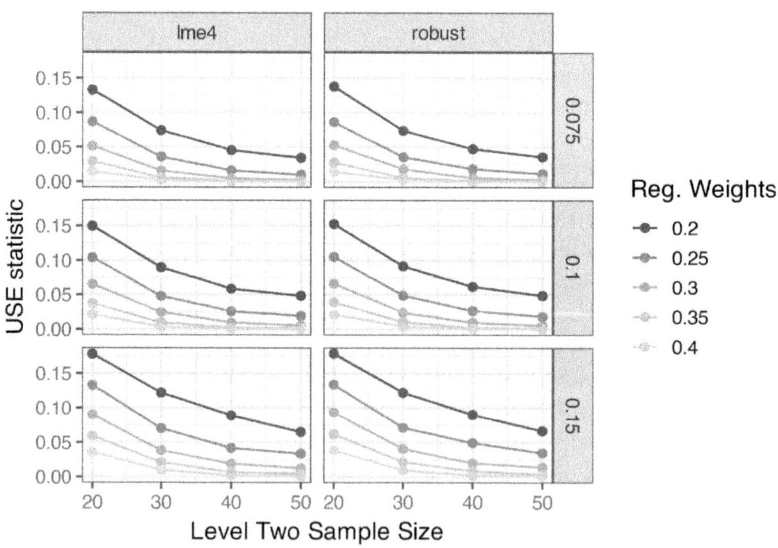

FIGURE 12.8 USE statistic estimates by level two sample size, regression weights, random effect variance (right panels), and model fitted (top panels).

population specified effect size is the smallest, 0.20, where USE statistics can approach 0.15 or 15% of the replications considered flipped signs. Most of the rest of the conditions were under 10% and even more were less than 5%, which would indicate a good amount of precision in the study design from the specified sample sizes.

When putting the USE and MSES statistics together, the narrative seems clear, for the sample sizes studied, the population effect size would need to be 0.30 or larger for the design to be powered strongly to balance detecting practically useful effects in tandem with minimizing risk of erroneously achieving a sample that would flip signs. If this effect size would be too large and there is a desire to detect the practically useful effect size as small as 0.20, then additional simulations would need to be run that increase the sample sizes to increase the precision to detect this magnitude of effect.

12.3.2.2 Treatment Effect Interaction

One other element that would be useful to explore is to evaluate the impact on power a treatment effect interaction may have. In this scenario, a treatment effect interaction would mean that the treatment effect is more or less effective depending on the pre-test score. For example, perhaps the treatment effect is stronger for those that started with higher vocabulary skills, this would be a positive treatment effect interaction. The contrast could also occur, there is a negative treatment effect interaction meaning that those that are lower would have larger treatment effects, perhaps due to having more ability to grow.

Interaction effects are more difficult to estimate well due to the estimates often being much smaller than the main effects. As such, estimating power and designing the study for estimating treatment interaction effects would be a good strategy to ensure properly powered studies. The following simulation formula specifies a treatment and pre-test interaction. When designing the simulation study to estimate power for the treatment effect and treatment by pre-test interaction, the following effect size values were crossed:

- Average Treatment Effect: 0.20, 0.25, 0.30, 0.35
- Treatment Effect Interaction: 0.05, 0.07

Furthermore, two competing models were explored:

- Linear Mixed Model
- Linear Model with Cluster Robust Standard Errors

The models with and without the treatment-pretest interaction effect were used to explore how sensitive the power estimates are to the magnitude of the interaction. For example, it would likely be true that the power estimates would be impacted when the treatment interaction is larger and ignored. The average treatment effect could be biased as well if the treatment interaction is not modeled when it is a part of the population model.

```
formula = knowledge ~ 1 + wj_pre + treatment + wj_pre:treatment
+ (1 | classroom)
```

First, the estimated parameters and their match to the simulation specified parameters should be done. This is not shown explicitly here, but on average the estimated parameters were close to the population simulation parameters. The parameters were plotted for both the treatment main effect as well as the treatment by pre-test score interaction effect.

Since the previous section only had the main effect, this section will focus on evaluating power for the interaction effect. Typically, if the study is properly powered to detect a practically useful interaction effect, then the main effect will also be properly powered. This is due to interaction effects being more difficult to estimate and are often smaller than main effects. There may be deviations from this general rule of thumb, so it is important to evaluate both to ensure that any terms being explored for adequate power are explored completely before conducting the study.

Figure 12.9 shows the MSES power estimates by the two models fitted (the mixed model and the robust model; top panels), the specified population interaction effect size (right panels), the level two sample size, and the two

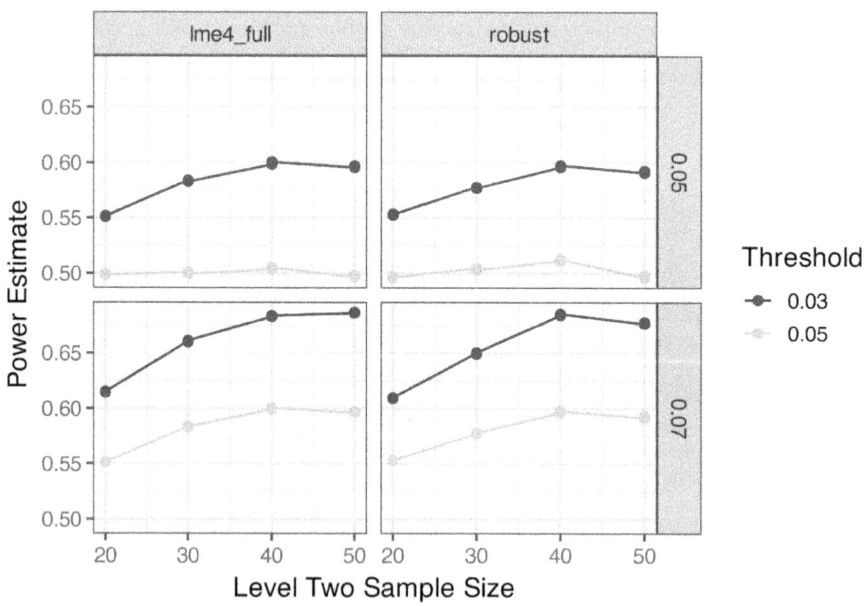

FIGURE 12.9 MSES statistics using the three specified thresholds for practically useful effect sizes for the interaction effect by level two sample size, regression weights (right panels), and the model fitted (top panels).

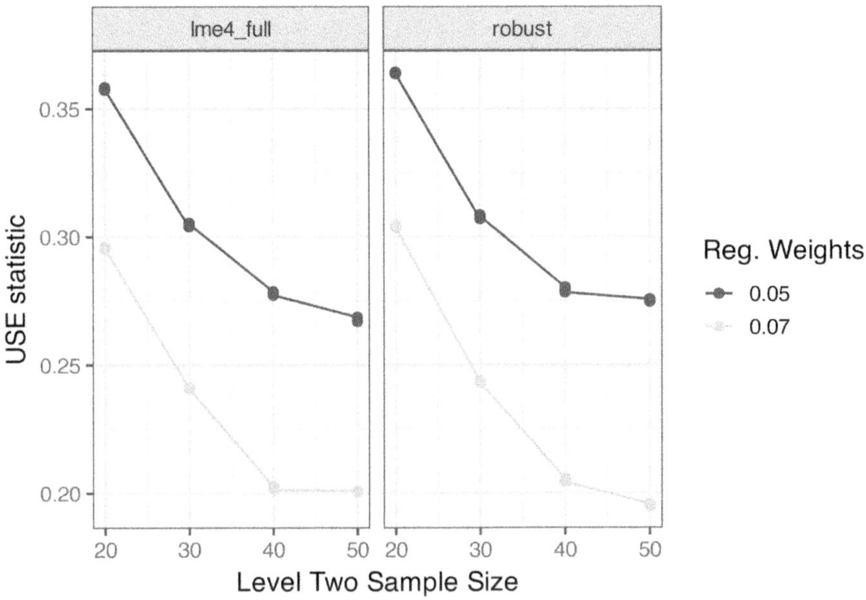

FIGURE 12.10 USE statistic estimates by level two sample size, regression weights, random effect variance (right panels), and model fitted (top panels) for the interaction effect.

practically useful thresholds. For the largest practically useful threshold identified, 0.05, the MSES statistics do not increase above 0.60, even for the most favorable data conditions (i.e., largest sample sizes), therefore, this would not be a favorable study condition to design a study around. Furthermore, for the smaller practically useful threshold, 0.03, this would be more viable, but not significantly so as the y-axis range needs to be considered here. The MSES statistics are still not greater than 0.70, but do get above 0.65. This would be a borderline condition and additional sample size requirements may be needed.

The USE statistics shown in Figure 12.10 show a similar story and further suggest that larger sample sizes would be needed for the simulation conditions studied to accurately detect the direction of the effect. In this situation, up to 35% of the simulated replications flipped signs. For strongly designed studies, this would be a characteristic that would not be attractive. The smallest USE statistics, about 20%, would represent a large percentage of replications that flipped signs. Ultimately, combining the USE and MSES information informs the study planning that additional simulation conditions should be explored to find a better mix of study conditions that can help lead to strong conclusions for this interaction term. If the interaction term is not of substantive interest, one could still explore if the simulated power statistics were strong for the main effect term in this situation. That is left up to the reader as an exercise.

13

Interrupted Time Series

13.1 Chapter Overview

This chapter introduces simulation methods for interrupted time series designs used to evaluate interventions implemented at a specific point in time. The chapter discusses modeling trends, level changes, and autocorrelation structures commonly encountered in time series data. Simulation examples demonstrate how design features such as series length, intervention timing, and noise structure affect power and estimation. The chapter emphasizes the role of simulation in understanding the strengths and limitations of interrupted time series designs under realistic conditions.

Interrupted time series (ITS), also called segmented regression analysis, is a within-subject design that can provide strong control for causal inference. The primary difficulty with ITS designs is that they require many observations or measurements; therefore, they can be challenging to implement when observations or measurements occur infrequently. However, ITS can be a design to consider for observations that have already been established or if observations can be obtained quickly and frequently.

In the simplest ITS design, a group of individuals is measured three to four times before a given treatment. These individuals are then measured on the same construct at least three to four times after the treatment. The desire to have at least three to four measurements before and after the treatment is given to establish the average location and trajectory both before and after the treatment. With fewer than three measurements, it is impossible to establish a longitudinal trajectory; even with three, only a linear trend can be established. Therefore, more measurements or observations are considered better for accurately establishing the trajectory before and after treatment. More measurements would be needed if there are seasonal or time-series-type trends that need to be taken into account.

The primary goal of the ITS design is to evaluate whether there is a change in mean level and/or a change in trajectory following the treatment. For example, suppose an exercise program was established for an individual to improve their health outcomes, such as blood pressure. In this study, participants may measure their blood pressure daily for 2 weeks before starting the new exercise

DOI: 10.1201/9781003177876-13

program. Then, they would start the new exercise program. During this exercise program, they would continue to measure their blood pressure for 2 months or, more generally, for a defined time frame that is long enough for the exercise to have an impact on their blood pressure. To evaluate how well the exercise program affects blood pressure in the study participants, the researchers could ask whether there are changes in blood pressure from baseline on day 1 of the exercise program. In addition, researchers could ask whether there is a change in blood pressure trajectory after the exercise program compared to the trajectory in the two weeks before starting the exercise program.

Imagining what the data from this design may look like, the initial period would, on average, be relatively stable blood pressure. There would be variation, but overall, the trend would be flat. Then, if the exercise helps reduce this person's blood pressure, you may see an initial drop once the exercise program starts. The treatment effectiveness could also show up without an initial mean change in blood pressure, but with a decrease in slope after the treatment, indicating that blood pressure is decreasing over time as the exercise program continues. The treatment could also combine the two types of effects, with a mean change initially and a slope change after the treatment. Finally, in this example, blood pressure would not drop forever; therefore, if the study were designed to measure individuals long enough, any potential decrease would flatten out to a new normal range. All of these types of effects can be estimated within an ITS design.

13.2 Adding Other Features to ITS Designs

ITS designs can become even stronger with additional features. For example, one could add a comparison group that does not receive the intervention, while still allowing for common changes at the intervention date. In this example, the baseline condition, also known as business as usual, would remain throughout the study. This could generalize to adding more than one treatment condition to evaluate the impact across conditions. This could be one way to study the impact of treatment dosage.

In addition to adding control groups or multiple treatments, design features such as crossover, withdrawal, reversal, or even multiple baseline designs can be used. Many of these design features come from single-case design, often used in low-incidence populations when rapid measurement is possible. Furthermore, these features require the trait to be measured to revert when the treatment is removed, which is not always possible. For example, once a student learns a new academic skill, it will likely revert only partially upon removal of the treatment. This could occur for behavior issues, however.

13.2.1 Crossover Designs

Crossover designs are those in which all individuals in the study receive the treatment, but are given it at different times. The name crossover comes from the idea that if visualizing the effect of the treatment and control groups, the outcome would cross over when the groups are given the treatment versus the control. Take, for example, a new treatment strategy for anxiety in individuals. Individuals are randomly assigned to receive the new treatment condition or the control (business as usual, i.e., the current/old treatment strategy or no treatment). After the initial baseline to establish anxiety symptoms in individuals, the new treatment is implemented for a period of time for those randomly selected. The other remains in the business-as-usual condition. Suppose that in this sequence, the anxiety symptoms of both groups are measured for a few months. Then, the crossover design is implemented, in which the treatment conditions are alternated across individuals. Those initially selected to receive the new treatment now flip to the business-as-usual condition, and those randomly assigned to the business-as-usual condition receive the new treatment. Similar to before, these individuals are again followed for a period of time to record their anxiety symptoms. Suppose the new treatment really is more effective. In that case, the anxiety symptoms of the two groups may cross if visualized using a line graph of the two groups, as the anxiety symptoms for those receiving the new treatment would decrease, and the business-as-usual condition may have higher anxiety symptoms.

In this type of design, it would also be able to estimate how long the new anxiety treatment would persist beyond when individuals are directly receiving it. This could be estimated by understanding how quickly, if at all, anxiety symptoms increase after the removal of the new treatment condition. That is, after the treatment is removed from the first group, how much, if at all, do the anxiety symptoms change? If there is very little change, the treatment strategies may persist in individuals. If the anxiety symptoms revert to the baseline level, it would suggest that the treatment is effective when implemented, but not effective once the treatment is stopped. In the **simglm** package, this can be represented by treating the treatment status as time-varying, with group-specific switch times.

13.2.2 Reversal / Withdrawal Designs

Withdrawal or reversal designs are similar to the crossover design discussed above but are more commonly found within the single-case experimental design literature (Smith, 2012). This is because these designs are typically discussed without a control group, so there is no crossover element. Instead, the treatment condition would be turned on/off at various intervals to understand the impact on the outcome. Staying with the anxiety example, imagine a design where the anxiety symptoms are collected at a baseline phase. Then, the treatment is implemented, and the anxiety symptoms are again measured for a period of time. The treatment is withdrawn for a period of time, then reinstated. This example is commonly referred to as an ABAB design, where A represents a period without the treatment, and B represents a period with the treatment. Additional periods to this structure can be added, but design standards suggest that the ABAB four-period cycle be the minimum required. Even though this is typically evaluated in the single-case experimental design literature without a control group, it is possible to add one if the population is large enough. This can be simulated by defining multiple post indicators (multiple knot locations), one for each on/off segment.

13.2.3 Multiple Baseline Designs

Multiple baseline designs are again commonly discussed alongside single-case designs (Slocum et al., 2022; Smith, 2012). Multiple baseline designs include varying lengths of baseline data collection across individuals. For example, baseline anxiety symptoms are collected for all individuals; then, a group is given the new treatment condition. The baseline data collection continues for the remaining individuals. Then, after the first group receives the new treatment condition, another group is given it a month later. At the same time, the rest who have not received the treatment are still in the baseline condition. This process continues until everyone is given the treatment condition, but in a highly staggered manner. This type of design is common in single-case experimental designs, as it helps ensure that the treatment effect is associated with the change in the outcome rather than an ecological or environmental factor. For example, imagine conducting a research study on anxiety that overlapped with the start of the COVID-19 pandemic. In this case, when the pandemic started, it would likely result in increased anxiety symptoms for many, if not all, individuals. In a multiple-baseline design, this effect would be present across the entire dataset and could be isolated. Similar to the above design features, this could be generalized to an ITS design with or without a control group. This corresponds to person-specific intervention times; in the **simglm** package, you can implement this by letting the knot location vary across individuals.

13.3 Simulating Data with ITS Designs Using simglm

One additional tool is needed to simulate an ITS design using a segmented regression approach. We will use a within-subject, or repeated-measures, approach to frame this discussion. Suppose we have a design with five measurement occasions before the treatment is implemented. The example below generates the time attribute as a sequence from -5 to 10, with a step size of 1. The negative values are the baseline, and the positive values are measured after the treatment is implemented. In the data-generation specification below, it is assumed that individuals grow by 0.5 units at each measurement occasion and that the intercept is 2. In this model, there is no direct impact of the treatment on the outcome at treatment implementation, and there is no change in the slope of the outcome.

```
library(simglm)

sim_args <- list(
    formula = y ~ 1  + time + (1 + time | id),
    fixed = list(time = list(var_type = 'time',
                             time_levels = seq(-5, 10, 1))),
    sample_size = list(level1 = 16, level2 = 50),
    randomeffect = list(int_id = list(variance = 5, var_level = 2),
                        time_id = list(variance = 1.5, var_level = 2)),
    error = list(variance = 10),
    reg_weights = c(2, .5)
  )

simulate_fixed(data = NULL, sim_args) |>
  simulate_randomeffect(sim_args) |>
  simulate_error(sim_args) |>
  generate_response(sim_args)
```

Differential slopes can be built into the population model via a segmented regression model at a specific treatment effect. The following model specification adds a new attribute named `treat_knot` This type of attribute is a dummy variable that indicates whether the treatment was active. More specifically, it is specified as an indicator attribute that equals 1 if the time attribute 0 or greater. In the **simglm** package, the knot argument is used to construct a post-intervention indicator (0 pre, 1 post) at the specified knot location.

Within this ITS framework, the treatment is assumed to be implemented between time $= -1$ and time $= 0$. Furthermore, an interaction between the linear slope and the attribute `treat_knot` indicates that the linear trend depends on whether the treatment is active. In the specification below, the treatment reduces the outcome on average by three units and decreases the linear slope by 0.15 units.

```
sim_args <- list(
    formula = y ~ 1  + time + treat_knot + time:treat_knot +(
      1 + time | id),
    fixed = list(time = list(var_type = 'time',
                             time_levels = seq(-5, 10, 1))),
    knot = list(treat_knot = list(variable = 'time',
                                  knot_locations = 0)),
    sample_size = list(level1 = 16, level2 = 50),
    randomeffect = list(int_id = list(variance = 5, var_level = 2),
```

```
                  time_id = list(variance = 1.5, var_level = 2)),
    error = list(variance = 10),
    reg_weights = c(2, .5, -3, -0.15)
  )

simulate_fixed(data = NULL, sim_args) |>
  simulate_randomeffect(sim_args) |>
  simulate_error(sim_args) |>
  generate_response(sim_args)
```

13.3.1 Adding a Control Group

In addition to adding the segmented regression parameters, a control group can be added and allowed to interact with them. In this case, the control group would be those who did not receive the treatment.

```
sim_args <- list(
    formula = y ~ 1  + time +
      control + treat_knot +
      time:treat_knot + control:treat_knot + (1 + time | id),
    fixed = list(time = list(var_type = 'time',
                             time_levels = seq(-5, 10, 1)),
                 control = list(var_type = 'factor',
                                levels = c('Control', 'Treatment'),
                                var_level = 2)),
    knot = list(treat_knot = list(variable = 'time',
                                  knot_locations = 0)),
    sample_size = list(level1 = 16, level2 = 150),
    randomeffect = list(int_id = list(variance = 5, var_level = 2),
                  time_id = list(variance = 1.5, var_level = 2)),
    error = list(variance = 10),
    reg_weights = c(2, .5, 3, -3, -0.15, 0)
  )

simulate_fixed(data = NULL, sim_args) |>
  simulate_randomeffect(sim_args) |>
  simulate_error(sim_args) |>
  generate_response(sim_args)
```

The model for the control group segmented data regression was specified as follows:

$$\begin{aligned} Y = \beta_0 + \beta_1\,\text{time} + \beta_2\,G + \beta_3\,\text{treat} \\ + \beta_4\,\text{time} : \text{treat} + \beta_5\,G : \text{treat} + \beta_6\,G : \text{treat} : \text{time} \\ + b_{0i} + b_{1i}\,\text{time} + \epsilon_i \end{aligned} \tag{13.1}$$

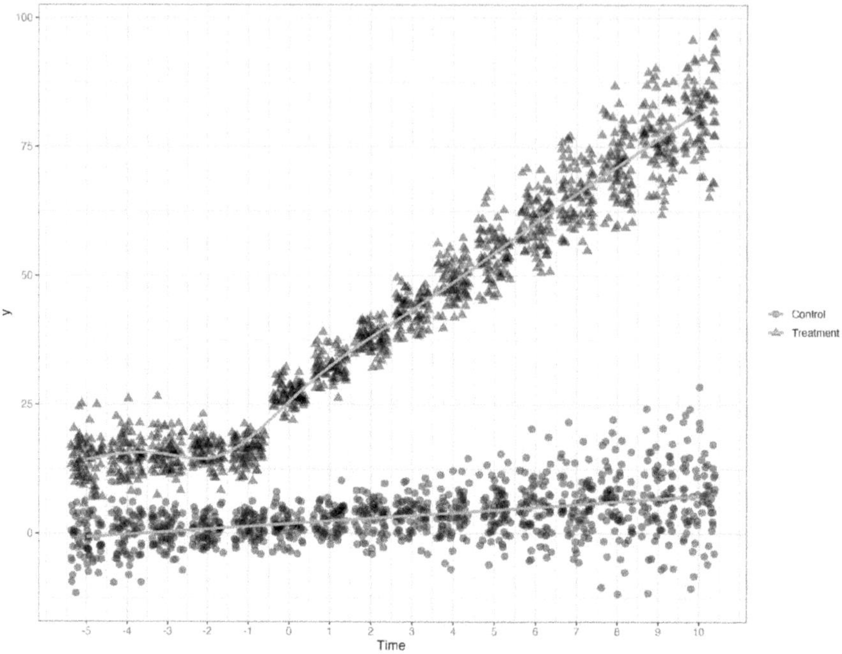

FIGURE 13.1 Interrupted time series design with a control group. The terms of the segmented regression approach underlying the simulation design are also depicted.

where *time* is centered when the treatment starts; that is, the treatment would be implemented between the -1 and 0 measurement occasions in Figure 13.1. The G term is a dichotomous attribute that equals 1 for those individuals in the treatment condition and 0 otherwise. The *treat* term is a dichotomous attribute that equals 1 for measurement occasions after the treatment (i.e., time 0 or greater), 0 otherwise. The b_{0i} and b_{1i} terms are random effects for the intercept and time, respectively, allowing the intercept and time slope to vary across individuals.

In this model, the beta terms would have the approximate interpretations:

- β_0: This is the mean level of the outcome for the reference group, time $= 0$ (at the time of treatment starting), and the control group.
- β_1: This is the common pre-treatment time trend shared by both groups.
- β_2: This is the average difference between the treatment and control group over the baseline phase evaluated at time $= 0$.

- β_3: This is the immediate mean change at the intervention for the control group (common shock or history effect).
- β_4: This is the average slope change in the time trend for the control group following the intervention.
- β_5: This is the additional immediate mean change for the treatment group at the intervention (i.e., time = 0), relative to the control group.
- β_6: This is the additional post-intervention slope change for the treatment group, relative to the control group.

These regression parameters are depicted in Figure 13.1 to tie the model equation to a visual. Often, the slope change and the initial treatment effect are of greatest interest when evaluating the treatment's effectiveness (i.e., β_5 and β_6). When there is no control group, any terms associated with the attribute 'control' would be removed from the model (i.e., β_2, β_5, and β_6). As such, the terms of interest would most likely be the treatment effect (β_3) and the change in slope (β_4).

13.3.2 Including More Than One Baseline

Simulating multiple-knot locations is possible using a segmented regression approach. This could be a case where multiple baselines occur, where different treatment dosages are offered, or where the treatment is also implemented to individuals at different time points. This approach could also be used to specify a non-linear trajectory via multiple segments. The trajectory within segments could be assumed to be linear or not.

The example below generates data with two treatment groups. In treatment group 1, the treatment will be implemented first, and the second treatment group will receive it later. Figure 13.2 shows how this is with simulated data. The first treatment group starts receiving the treatment at time 0, centered so that the intercept is more interpretable. Then, the second treatment group receives the treatment at time 4. In this case, the data generation has two knot attributes, one for each of the two treatment groups. This allows the indicator attributes to line up properly with the assumed design.

This design can be further extended to more knot locations; therefore, there is no functional limit to the number that can be generated or simulated. To simulate more, additional terms are specified within the knot simulation argument. The real limit would be what could be realistically implemented in a single study. A good point to remember is that, although we can simulate it, the data should reflect the study being planned.

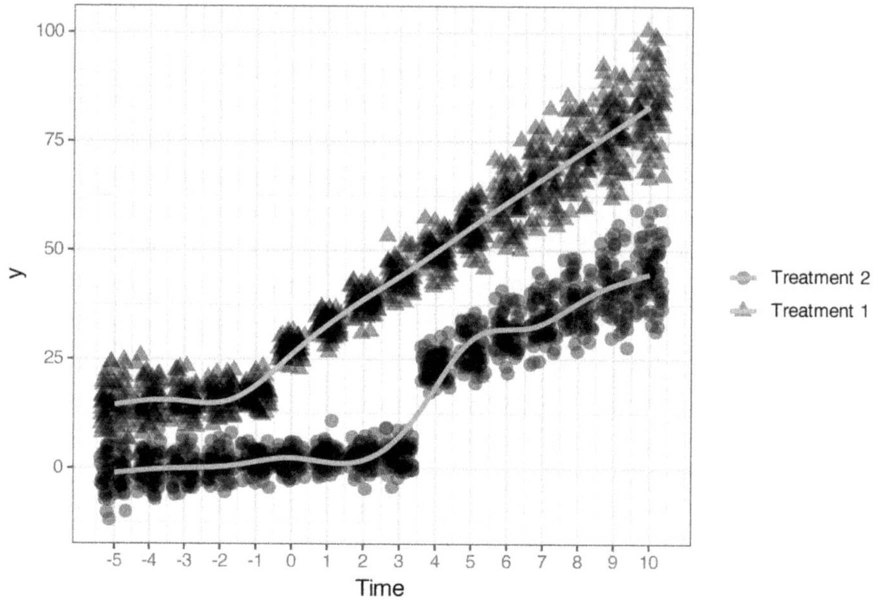

FIGURE 13.2 Interrupted time series design that has two treatment groups that are given the treatment at different times.

13.4 Evaluating Power for ITS Design

Estimating power for an ITS design offers many options to build into the framework. For example, the number of baseline and post-treatment measurement occasions needs to be considered. Whether there is a control group, or whether groups receive the treatment at different times, affects how precisely the treatment effect can be estimated. Other traditional mechanisms also needed to be explored; for example, the number of units to include, the amount of missing data, the hypothesized effect size, and, if doing simulation-based power analyses, the practically relevant effect size thresholds.

To tradeoff some of these considerations, the following study designs were considered for this hypothetical example.

- Number of measurement occasions: 10 vs 16
- Number of individuals: 150, 250, 400
- Treatment Effect (mean change): 0.1, 0.15, 0.2
- Treatment Slope Change: 0.05, 0.1

These conditions were fully crossed and explored. To facilitate interpretation, the results for the 16 measurement occasions will be examined first, followed by those for the 10 measurement occasions. For a full analysis, exploring other effects, such as missing data, is important and is left as an exercise for the reader. In longitudinal type designs like this, dropout missing data, where an individual starts the study but does not complete all occasions, is a common missing data mechanism that would be important to explore the impact on the power estimates.

```
sim_design <- list(
    formula = y ~ 1  + time +
      control + treat_knot +
      time:treat_knot + control:treat_knot + control:time:treat_knot +
      (1 + time | id),
    fixed = list(time = list(var_type = 'time',
                        time_levels = seq(-5, 10, 1)),
                 control = list(var_type = 'factor',
                                levels = c('Control', 'Treatment'),
                                var_level = 2)),
    knot = list(treat_knot = list(variable = 'time',
                                  knot_locations = 0)),
    randomeffect = list(int_id = list(variance = 1, var_level = 2),
                        time_id = list(variance = .25, var_level = 2)),
    model_fit = list(formula = y ~ 1  + time +
      control + treat_knot +
      time:treat_knot + control:treat_knot + control:time:treat_knot +
        (1 + time | id),
          model_function = lme4::lmer),
    error = list(variance = 1),
    replications = 2500,
    extract_coefficients = TRUE,
     power = list(
    thresholds = list('controlTreatment:treat_knot' = c(.05, .075, .1, .15),
                 'time:controlTreatment:treat_knot' = c(.05, .075, .1, .15)),
    type_use_sign = list('controlTreatment:treat_knot' = 'positive',
                       'time:controlTreatment:treat_knot' = 'positive')
  ),
    vary_arguments = list(
    reg_weights = list(
      c(0, .1, 0, 0, 0, 0.1, 0.05),
      c(0, .1, 0, 0, 0, 0.15, 0.05),
      c(0, .1, 0, 0, 0, 0.2, 0.05),
      c(0, .1, 0, 0, 0, 0.1, 0.1),
      c(0, .1, 0, 0, 0, 0.15, 0.1),
      c(0, .1, 0, 0, 0, 0.2, 0.1)
    ),
    sample_size = list(
      list(level1 = 16, level2 = 150),
      list(level1 = 16, level2 = 250),
      list(level1 = 16, level2 = 400)
    )
  )
  )

its_sim <- replicate_simulation(sim_design)
```

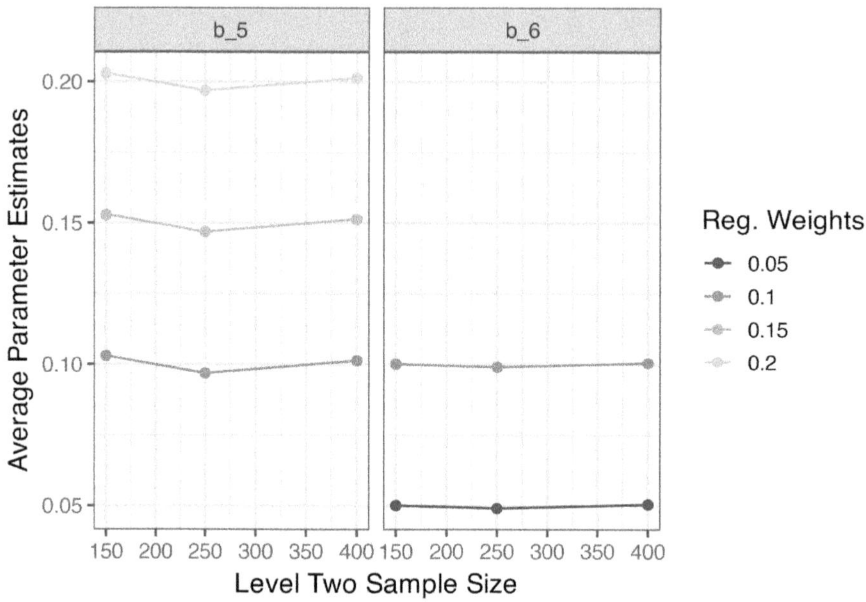

FIGURE 13.3 Average simulated model estimates by level two sample size, term, and regression weights.

First, confirming that the parameters were estimated well. This can be done by comparing average estimates to the population regression weights specified in the simulation design, or visually. Figure 13.3 shows this for the design. For this figure, an ideal situation would be for the average estimates to form a straight line right at the population regression weight parameters. This is the pattern discerned for the two primary terms of interest that were varied within the simulation design; therefore, continuing with the power analysis is possible.

Figure 13.4 shows the MSES power estimates from this simulation design. The left panel contains the average mean change in the treatment effect (i.e., β_5) and the right panel shows the average slope change for the treatment group after the treatment (i.e., β_6). The specified population effect sizes are shown in the right-side panels; some panels lack data because those effect sizes were not specified for those terms. The number of measurement occasions remained fixed in this example; those are not shown.

Overall, the MSES power estimates are strong for most conditions for the average mean change (β_5). In most conditions and across the threshold values, the estimated MSES values were greater than 0.5, with many above 0.75. The estimated MSES was not as strong for estimating the slope change, however. This is due to the population slopes being smaller than the mean change, a

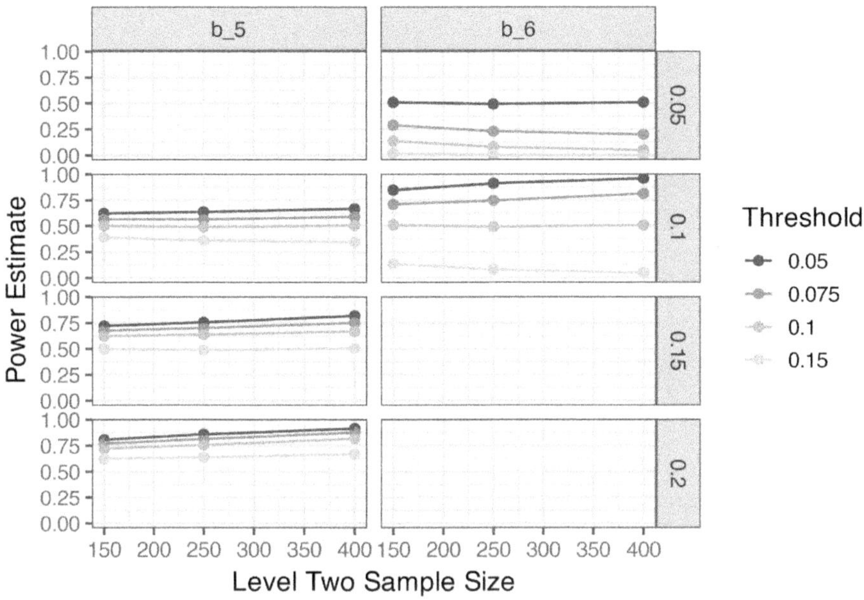

FIGURE 13.4 MSES power estimates using the four specified thresholds for practically useful effect sizes by level two sample size, term, and regression weights.

phenomenon I have observed in my experience. Powering a study for detecting a slope change would likely increase the power to detect a mean change.

To fully understand the implications of the estimated power, evaluating the USE statistics is often a good idea. Ensuring that the estimated effect can be detected as either positive or negative is a good minimum to identify for follow-up studies. Figure 13.5 shows the estimated USE statistics for this simulation design. The panels show the same terms as in the previous figure, and the lines show the specified population effect sizes for each term. The USE statistics are large for the smallest two-level two-sample sizes for many of the specified population effect sizes. For the largest level two sample size, the USE statistics are less than 10% for the larger two population effect sizes in each case.

When exploring the MSES and USE statistics in tandem, level two sample sizes of 400 would likely be needed to have adequate power to estimate effects with correct directionality and with practically relevant effect sizes. Further exploration of additional individuals would be warranted, especially given that the current structure assumes no missing data across these individuals, which is highly unlikely given this many measurement occasions.

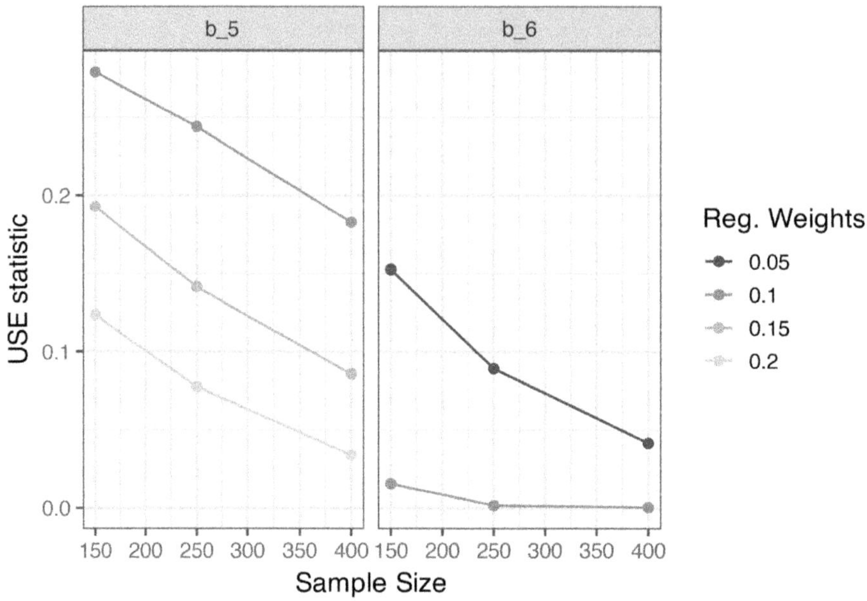

FIGURE 13.5 USE statistic estimates by level two sample size, term, and regression weights.

13.4.1 Smaller Time-Series Sequences

```
sim_design <- list(
  formula = y ~ 1  + time +
    control + treat_knot +
    time:treat_knot + control:treat_knot + control:time:treat_knot +
    (1 + time | id),
  fixed = list(time = list(var_type = 'time',
                           time_levels = seq(-3, 6, 1)),
               control = list(var_type = 'factor',
                              levels = c('Control', 'Treatment'),
                              var_level = 2)),
  knot = list(treat_knot = list(variable = 'time',
                                knot_locations = 0)),
  randomeffect = list(int_id = list(variance = 1, var_level = 2),
                      time_id = list(variance = .25, var_level = 2)),
  model_fit = list(formula = y ~ 1  + time +
                     control + treat_knot +
                     time:treat_knot + control:treat_knot +
                     control:time:treat_knot + (1 + time | id),
```

```r
                    model_function = lme4::lmer),
  error = list(variance = 1),
  replications = 2500,
  extract_coefficients = TRUE,
  power = list(
    thresholds = list(
      'controlTreatment:treat_knot' = c(.05, .075, .1, .15),
      'time:controlTreatment:treat_knot' = c(.05, .075, .1, .15)),
    type_use_sign = list(
      'controlTreatment:treat_knot' = 'positive',
      'time:controlTreatment:treat_knot' = 'positive')
  ),
  vary_arguments = list(
    reg_weights = list(
      c(0, .1, 0, 0, 0, 0.1, 0.05),
      c(0, .1, 0, 0, 0, 0.15, 0.05),
      c(0, .1, 0, 0, 0, 0.2, 0.05),
      c(0, .1, 0, 0, 0, 0.1, 0.1),
      c(0, .1, 0, 0, 0, 0.15, 0.1),
      c(0, .1, 0, 0, 0, 0.2, 0.1)
    ),
    sample_size = list(
      list(level1 = 10, level2 = 150),
      list(level1 = 10, level2 = 250),
      list(level1 = 10, level2 = 400)
    )
  )
)

its_sim_small <- replicate_simulation(sim_design)
```

When exploring a smaller number of measurement occasions (10 instead of 16), only the estimated MSES are shown. The average estimates were very close to the specified population effect sizes. The USE statistics were also similar to those just shown, but were elevated comparatively. This further suggests that more data, not less, is needed, as shown in this example. Figure 13.6 shows the estimated MSES statistics across this study design. Overall, there are only subtle differences compared to the longer number of measurement occasions, although any change would result in the estimated MSES being smaller than in the previous example. This aligns with experience: the number of level two units tends to have a greater impact on estimated power than the number of level one units. Within the ITS framework, however, it is important to ensure there are sufficient measurement occasions before and after the treatment to estimate the slope accurately.

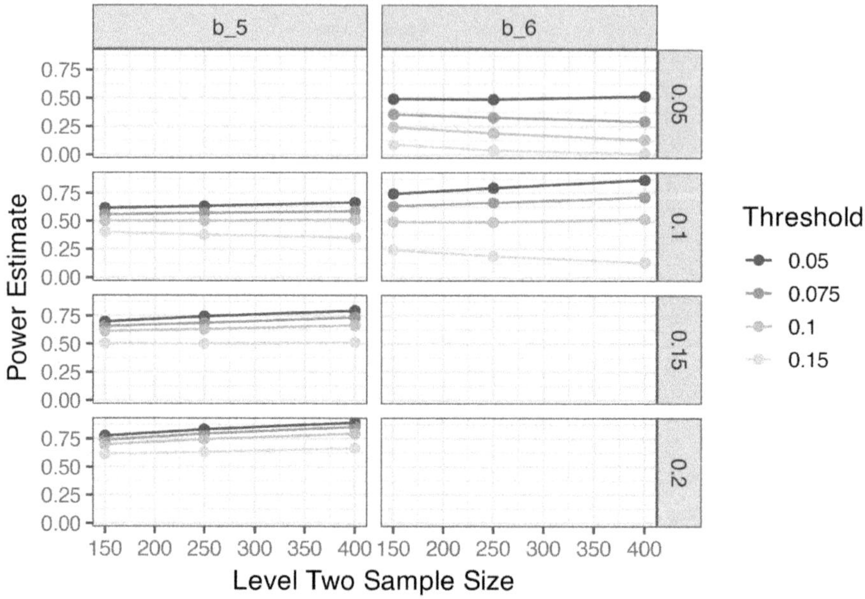

FIGURE 13.6 Alternative power estimates using the four specified thresholds for practically useful effect sizes by level two sample size, term, and regression weights.

All of these explorations suggest that a larger level two sample size would be warranted; however, the number of measurement occasions per individual would not need to be altered unless driven by theory. Further, exploring the impact of missing data on estimated power would be important, especially for dropout-type missing data, where individuals are more likely to drop out of the study as the study length increases. This may not have a strong impact on the estimated power for the mean change at the time of treatment implementation, as many individuals may remain in the study for at least that long. It could adversely affect the estimated slope change after treatment, as there would likely be less data to estimate the post-treatment slope as confidently. Exploring seasonality, autocorrelation, and other design features that can impact power estimates would also be good.

14

Regression Discontinuity

14.1 Chapter Overview

This chapter applies simulation-based methods to regression discontinuity designs. The chapter discusses the role of the running variable, cutoff specification, and functional form assumptions. Simulation examples illustrate how bandwidth selection, sample size near the cutoff, and treatment effect heterogeneity influence estimation and power. By simulating data under known conditions, the chapter shows how researchers can evaluate the plausibility and performance of regression discontinuity analyses before implementation.

Regression discontinuity design (RDD) is a quasi-experimental design aimed at making causal inferences from observational data. In this method, a cutoff is determined to identify if an individual is in the treatment or control group. The local average treatment effect can be estimated using data just above and just below this cutoff score. Under standard assumptions (e.g., continuity and no precise manipulation at the cutoff), RDD can yield strong internal validity for local causal effects near the cutoff.

To give a concrete example of this design in practice, let us look at a study by Jacob and Lefgren (2004) that explored the impact of a policy decision in Chicago Public Schools in 1996 that, based on standardized achievement tests, guided summer school and other remedial education practices. Essentially, students were enrolled in summer school or some form of remedial education if they did not meet the defined standardized achievement scores in reading and math in grades 3, 6, and 8. The defined standardized achievement scores were roughly equivalent to the 20th percentile for each grade level. Ultimately, the study found that those receiving remedial education in third grade showed strong achievement gains, but this was not the case for those in sixth grade.

The RDD is often estimated using linear regression or non-parametric methods. We can do either with the **simglm** package when estimating power. However, this book will focus on the parametric approaches to estimation using ordinary least squares and robust methods. Those interested in reading more about RDD are directed toward Cunningham (2021) and Huntington-Klein (2021).

DOI: 10.1201/9781003177876-14

14.2 RDD Terminology

To start discussing simulation of data and estimating power from RDD designs, it is important to consider some key terms that are commonly used within the RDD literature. Defining these terms will aid in discussing and simulating the RDD model.

- **Running Attribute** is the data attribute for which the determination of the treatment occurs. Often, there is a single attribute defined as the running attribute. However, this will only fit some situations. Sometimes, more than one running attribute determines whether the treatment is activated.
- **Cutoff** is the point in the running attribute where the treatment occurs. This is often a hard binary cutoff (i.e., on/off, yes/no). However, this can be non-binary as well. Furthermore, it is possible to have multiple cutoffs as well. The focus of this book will be a single cutoff.
- **Bandwidth** is the window around the cutoff used for estimation; smaller bandwidths reduce functional-form bias but increase sampling variability.

These are the primary terms to remember when discussing the RDD design. However, within the RDD literature, there are two general types as well. First, *sharp RDD studies* are those that were described above where a hard binary cutoff is used. For example, those above a specific value of the running attribute are given the treatment, and those below are not. In contrast, *fuzzy RDD studies* are those where the cutoff is not absolute. Instead, the treatment's probability increases as the running attribute increases.

There is a sub-design within each of the two general types of RDD studies that can be considered. This has been termed the regression kink design (Card et al., 2015). This type of RDD study can be used where there is no jump (either up or down) at the cutoff score. Instead, there is a slope change after the cut-off. As simulating data from an RDD, the mechanics are similar to that of an ITS design using knot attributes. Much of the mechanics were shown in Chapter 13.

14.3 Simulating Data from RDD

To show how to simulate data from an RDD, a study by Duchini (2017) will frame the process. This study explored if college remediation increases persistence (i.e., continued enrollment in college) and performance (i.e., doing well in college coursework). This study used data from the Italian tertiary education system; the sample data used was just under 3,000 students. In

2009 and 2010, an economics department in Italy implemented an entrance exam using a standardized test that had three subsections: math, verbal comprehension, and logic. Admission decisions were made using a weighted average of the three subsections. Remedial coursework was determined solely based on the math score.

The study used a sharp RDD where the cutoff was a specific score on the math entrance exam. Those below the cutoff but still admitted to the program were assigned remedial math courses; those above the cutoff and admitted did not take the remedial math course.

14.3.1 The Sharp RDD Model Used

The paper used two sharp RDD models to explore the impact of the remedial program and future program performance. First, the author used local linear regression with the following linear model:

$$\begin{aligned} Y_i = & \beta_0 + \beta_1 D_i + \beta_2 \, \text{mathscore}_i \\ & + \beta_3 \, D_i : \text{mathscore}_i + \beta_c \, \text{controls}_i + \beta_{ch} \, \text{cohort}_i + \epsilon_i \end{aligned} \tag{14.1}$$

Equation @ref{eq:localrdd} includes the primary treatment effect, D_i, which would estimate the average treatment effect at the cut-score for those who received the remedial coursework. To visualize the outcome, this would be the increase or decrease separating students just below compared to those just above the cut-score. The math scores were centered on the cut-score. Therefore, the intercept would represent the average outcome score for those not in remediation. There was also an interaction term that estimated if there was a slope change for those in remedial coursework. Finally, the author also statistically adjusted for various other features like student sex, immigration status, type of high school attended, performance in high school math, and for the different cohorts.

The author implemented local linear regression using varied bandwidth values from the cut-score. The bandwidth values implemented were those within 1.25, 2.5, or 5. This means that the data were filtered before estimating the model in Equation (14.1) such that scores were within 1.25, 2.5, or 5 units from the math cut-score.

Finally, the authors also explored a polynomial regression model with a quadratic form of the math scores. This new model looked like this:

$$\begin{aligned} Y_i = & \beta_0 + \beta_1 D_i + \beta_2 \, \text{mathscore}_i + \beta_3 \, \text{mathscore}_i^2 \\ & + \beta_4 \, D_i : \text{mathscore}_i + \beta_5 \, D_i : \text{mathscore}_i^2 \\ & + \beta_c \, \text{controls}_i + \beta_{ch} \, \text{cohort}_i + \epsilon_i \end{aligned} \tag{14.2}$$

This allows for flexibility in the functional association between the math score and the outcome. Each of these two modeling forms will be shown how to simulate.

14.3.2 Simulating RDD Data

When simulating the RDD data, the model described above will be the primary model simulated. For the example, the number of first year credits, one of the primary outcomes for the study by Duchini (2017) will be used. When establishing the regression coefficients, Table 8 from Duchini (2017) shows the estimated coefficients for the number of first year credits. Depending on the bandwidth used, the associated effect was from one to three credits fewer for those that were below the math entrance exam cut-score and placed in a remedial math course.

The model formula simulated in the first instance is: `formula = num_credits ~ 1 + math_score + D_post + math_score:D_post`. The model formula matches the first model equation above in which the math score is the assignment attribute and the D attribute is equal to one for those who received the remediation math course and zero for those that did not. Finally, an interaction between the remediation status and the math score is shown which allows for the relationships between number of credits and the math scores to be different based on those that enroll in the remediation course compared to those that do not.

Following the formula specification, the simulation arguments for those attributes is shown the code chunk below. For this model, the fixed and post simulation arguments are needed. This is needed as the treatment effect is defined as a function of the math score, more specifically a cut-score with the math scores. The math score was defined to have a mean of -2.78 and a standard deviation of 3.65, this was described in the article by Duchini (2017) when describing Figure 1 of their article. Finally, the treatment effect, named "D_post", in the formula code, is created based on a function of the simulated math scores. This attribute must be specified using post processing, which is why the attribute has the "_post" suffix. The D attribute will equal one, enroll in the remedial math course, when math scores that are less than or equal to zero; otherwise, the D attribute will equal zero, not enroll in the remedial math course, when math scores are greater than 0.

For the other simulation conditions needed, the sample size was specified to be 2,785, the same as the primary study. The error variance was specified as equal to five, this value was not directly specified within the study; therefore, this would need to have some trial and error to identify the appropriate metric. Finally, the population regression coefficients were specified as: 28 (intercept), 0.47 (math score slope), -1 (treatment effect), and -0.15 (slope difference for the math score slope for the remedial group).

TABLE 14.1 First five rows of the simulated regression discontinuity data.

Int.	Math	D Post	Math:D Post	ID	Err.	Fixed	Random	Num. Cred
1	−6.234	1	−6.234	1	3.559	25.00	0	28.56
1	7.344	0	0.000	2	4.942	31.45	0	36.39
1	−3.456	1	−3.456	3	−1.599	25.89	0	24.30
1	−5.033	1	−5.033	4	−1.169	25.39	0	24.22
1	−0.465	1	−0.465	5	2.148	26.85	0	29.00

```
fixed = list(
  math_score = list(var_type = 'continuous', mean = -2.78, sd = 3.65)
)
post = list(
  D_post = list(variable = 'math_score',
                fun = 'ifelse',
                condition = '<= 0',
                yes = 1,
                no = 0)
)
```

Table 14.1 shows the first five rows of the simulated data.

Figure 14.1 shows the simulated RDD data binning the 2,785 individuals (this was the sample size from Duchini (2017)) by binning the math scores into 24 equally sized bins and computing the average number of credits within each bin. The discontinuity is shown by the gaps in the lines at the math score equal to 0. This gap is subtle here as the average difference in the number of credits was one credit less for those in the remediation group at the cut-score of zero. There was a slope difference specified for the remediation group where that group has a flatter slope or a smaller association between the math score and the number of credits.

14.3.2.1 Specify a Non-Linear Relationship

As shown above, the study by Duchini (2017) also estimated a polynomial model where the relationship between the math scores and the number of credits was not linear. This data can also be simulated as the underlying data generating model with the **simglm** package. This could be of interest to explore the impact of functional form in estimating the treatment effect of interest. Furthermore, this could impact the power estimates and would be a good element to include when estimating the power. From a study design perspective, if it is possible to adequately plan for the most complicated scenario, then the study would also be adequately powered for simpler data conditions.

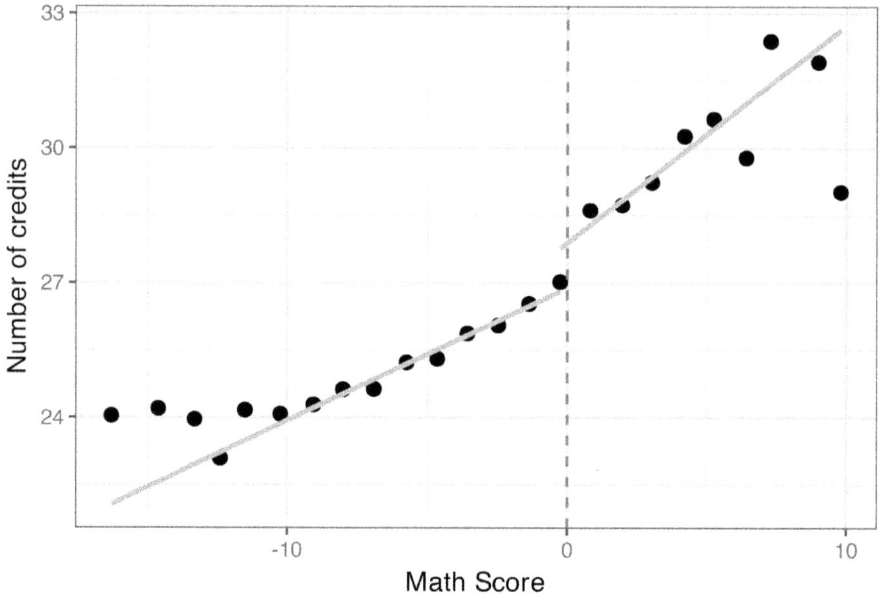

FIGURE 14.1 Average number of credits by binning the math scores into 24 evenly spread out bins.

When simulating data that are non-linear via a polynomial term, the `poly()` function is used for this. These can be specified within or with an interaction. For example, the model equation with a quadratic term and all interactions could be specified as: `num_credits ~ 1 + poly(math_score, degree = 2, raw = TRUE) * D_post`. To complement this, there would then be six regression coefficients representing the intercept, linear slope, quadratic slope, remedial course effect at the cut-score, slope difference for the remedial course group, and quadratic slope difference for the remedial course group. In the simulated example simulated, the regression coefficients were specified as: `reg_weights = c(28, 0.5, 0.01, -5, -0.4, -.01)`. These correspond to the terms described in words above.

Figure 14.2 shows the resulting structure of the simulated data with small quadratic terms. Similar to the earlier figure, not all the data are shown, instead 24 unique bins were created for the math score attribute and the average number of credits were computed within each of the bins. Notice that the quadratic effect is larger for those above the cut-score and the average quadratic effect for those in the remedial course was specified to be zero. The linear slope for the remedial group was also specified to be close to zero and the simulated RDD effect was specified as -5. This means that the average difference in the number of credits at the cut-score was specified to be five credits fewer for the remedial course group.

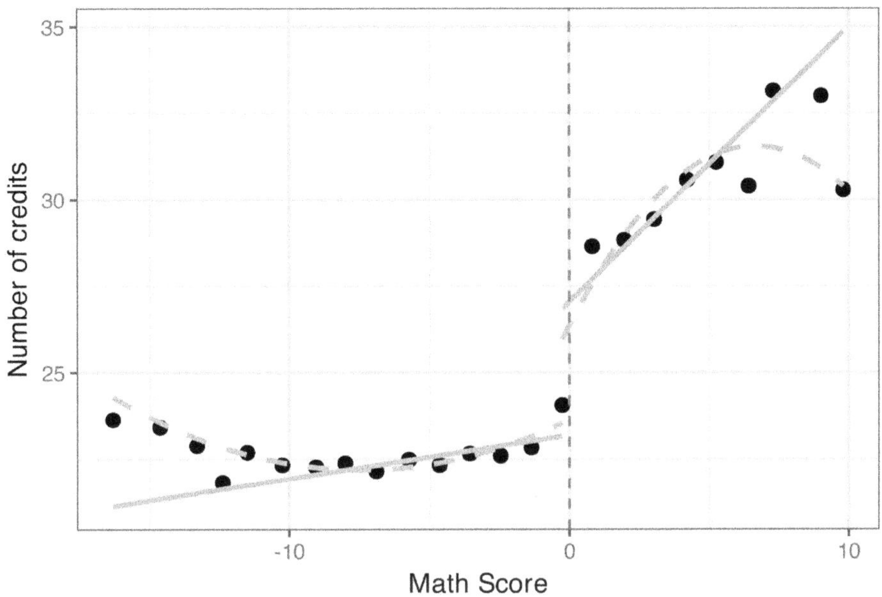

FIGURE 14.2 Average number of credits by binning the math scores into 24 evenly spread out bins.

14.4 Estimating Power from RDD

Study planning for an RDD study would follow a similar structure to the previous chapters. Identifying ranges of likely and impactful effect sizes for the RDD effect would be of primary interest. Furthermore, hypothesizing the amount of error, form of change (i.e., linear or non-linear), varying the sample size, or other study design implications would need to be considered prior to doing the simulation design. For this example of estimating power, the likelihood of dropping out after the first year from the study by Duchini (2017) will be used to frame the analysis. The estimated RDD treatment effect ranged from 0.01 to 0.07 across the variety of RDD models considered by Duchini (2017). The following conditions will be varied in the simulation design for estimating power.

- Effect Size: 0.05, 0.075, 0.1, 0.15
- Sample Size: 500, 750, 1000, 1500
- Fitted Model: Interaction or Additive

The primary fixed effect for the D_post term will be of primary interest to explore. In the models that follow, the effect of interest is in the log-odds scale (i.e., the coefficient for the D_post term).

The average parameter estimates were compared to the population effect sizes specified through the simulation design. The additive model that does not include the interaction term tends to overestimate the effect size, whereas the interaction model is close to the true population effect sizes specified. Given the biased nature of the additive model, these parameters would not be of interest, but we will continue with the example to show the difference in the estimated MSES statistics from the two models.

Figure 14.3 shows the estimated MSES statistics from the two model frameworks by the four thresholds defined for practically relevant effect sizes. As expected given they had larger parameter estimates, the MSES is larger (inflated) for the additive only model compared to the interaction model. Overall, the MSES statistics are modest for the interaction model, suggesting that larger sample sizes would need to be explored for a complete understanding of study planning.

The USE statistics shown in Figure 14.4 further show the problem with the current simulation conditions and their inadequacy for a real study. The USE statistics are approached 50% in many of the conditions. This means that

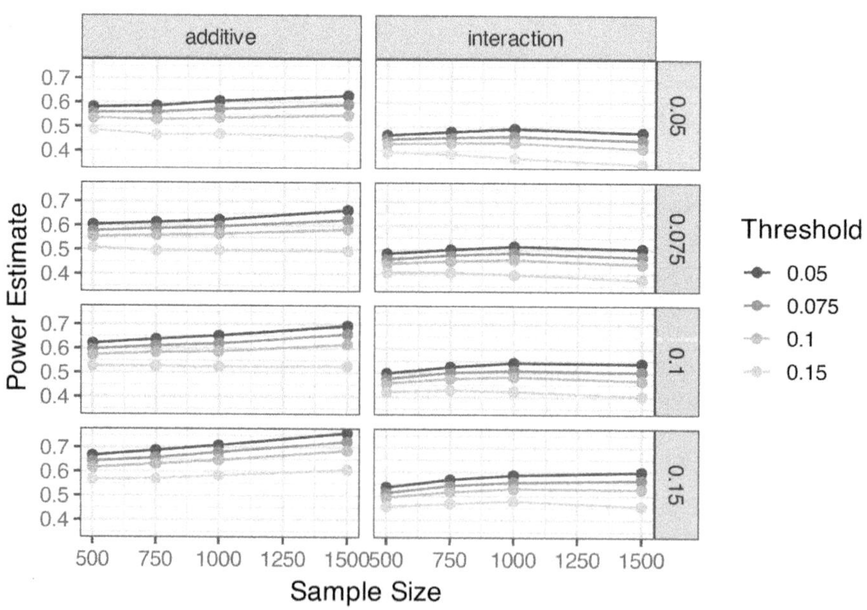

FIGURE 14.3 MSES statistics using the four specified thresholds for practically useful effect sizes by level two sample size, term, and regression weights.

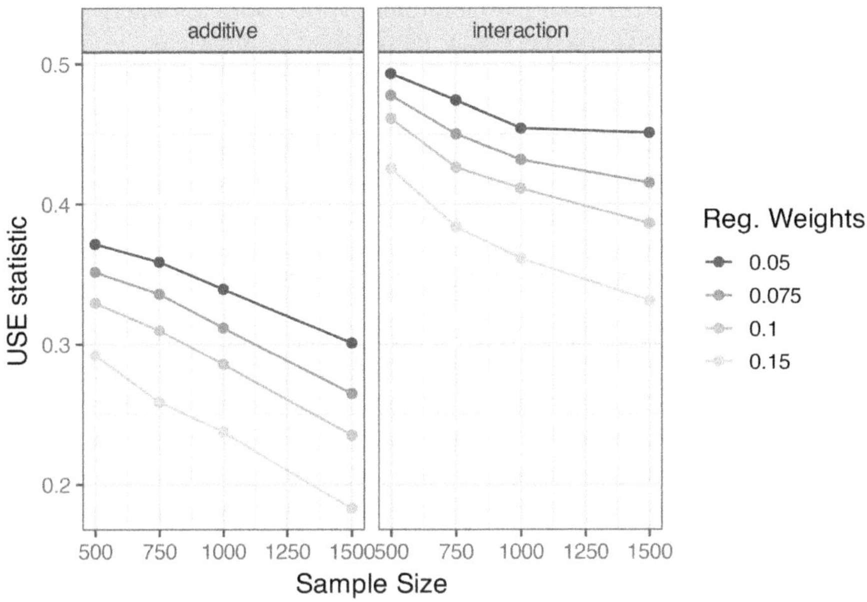

FIGURE 14.4 USE statistic estimates by level two sample size, term, and regression weights.

almost 50% of the parameter estimates are being estimated in the opposite direction as the population effect sizes specified. USE statistics near 50% indicate the estimate is highly unstable at these sample sizes under the assumed data-generating process, suggesting the design is underpowered for reliable directional conclusions.

14.5 Extensions

14.5.1 Non-Linear Trajectories

Although many introductory treatments of regression discontinuity rely on linear approximations near the cutoff, the underlying relationship between the running variable and the outcome may be non-linear even within relatively narrow bandwidths. When this is the case, assuming linear trajectories on either side of the cutoff can lead to biased estimates of the discontinuity and misleading conclusions about both effect size and uncertainty. Non-linear structure may arise from ceiling or floor effects, saturation in learning or performance outcomes, or institutional rules that induce curvature in the response to the running variable. As a result, study planning for RDD designs should consider whether plausible non-linear trajectories are likely near the cutoff and how sensitive inference may be to those assumptions.

Simulation provides a particularly useful framework for exploring these issues prior to data collection. By specifying non-linear data-generating mechanisms, such as quadratic or higher-order polynomial relationships, or differential curvature across treatment conditions, researchers can evaluate how alternative model specifications affect bias, error rates, and power to detect practically meaningful effects. Importantly, accurately estimating non-linear trajectories typically requires larger sample sizes near the cutoff than linear models, as additional parameters must be estimated and curvature is identified through relatively subtle changes in the outcome across the running variable. When sample sizes are limited, quadratic terms may be weakly identified, leading to unstable estimates, inflated uncertainty, or increased Type S and Type M errors. From a design perspective, planning under more complex non-linear scenarios can help ensure robustness: if a study is adequately powered under conservative assumptions about curvature and slope changes, it is likely to perform well under simpler conditions. This makes explicit the exploration of non-linear trajectories and their associated sample-size demands, which is an important component of rigorous RDD study planning and simulation-based power analysis.

14.5.2 Multi-Site and Clustered RDD Designs

Many applied regression discontinuity studies are conducted in settings where units are nested within higher-level structures, such as students within schools, applicants within institutions, or individuals within geographic regions. In these multi-site contexts, treatment assignment is still determined by a common cutoff on the running variable, but outcomes may be correlated within clusters and treatment effects may vary across sites. These features introduce additional design considerations beyond those illustrated in the single-level examples in this chapter.

From a study-planning perspective, clustered RDD designs are influenced not only by the total sample size near the cutoff, but also by the number of clusters, the distribution of units across clusters, and the degree of between-cluster heterogeneity. Analytically, these designs are commonly handled using cluster-robust standard errors or mixed-effects models that include random intercepts (and potentially random slopes) for clusters. The simulation framework demonstrated in this chapter can be naturally extended to these settings by incorporating random effects into the data-generating process and fitting appropriate clustered or multilevel models during estimation. While a full treatment of clustered RDD designs is beyond the scope of this chapter, careful attention to functional form, correct specification of the discontinuity, and sensitivity to modeling assumptions remain central principles when planning and evaluating multi-site RDD studies.

15

Difference in Differences

15.1 Chapter Overview

This chapter focuses on difference-in-differences designs for causal inference using observational or quasi-experimental data. The chapter discusses the parallel trends assumption and how violations affect estimation. Simulation examples demonstrate how pre-treatment trends, treatment timing, and noise structure influence results. The chapter emphasizes the use of simulation to assess robustness to assumption violations and to guide design and analysis choices in applied settings.

Difference-in-differences design (DID) is a popular causal inference method for observational data, sometimes referred to as quasi-experimental designs. DID estimates the average treatment effect by comparing the difference between two time points on the treatment group after adjusting for differences that would have occurred without the treatment. Adjusting for differences that would have occurred without the treatment is done by including a control group that does not receive the treatment. Including these effects helps to disentangle the treatment effect by removing group differences and time effects and any differential difference between the treatment group and control group over time is attributed to the treatment. This is where the name, difference-in-differences comes from.

One way that the DID design differs from a RCT design stems from random assignment. In a RCT design, the treatment is assigned randomly to the study participants. In the DID design, random assignment either was not done or was not able to be done due to ethical or other concerns. Often, the DID design is implemented at a larger systematic level using administrative data where the groups are intact already with one receiving and one not receiving the treatment condition. These types of systematic questions have been used to study the impact of minimum wage on employment outcomes (Card and Krueger, 1994), the impact mandated maternity benefits have on employment outcomes (Gruber, 1994), or the impact of medicaid on mortality (Miller et al., 2021).

In the simplest case, the DID would include two time points, a time point before and a time point after, and also two groups, the one that received

DOI: 10.1201/9781003177876-15

the treatment and one that did not. This case is often referred to as a 2x2 DID design. The average treatment effect can be estimated with the following regression:

$$Y = \beta_0 + \beta_1 time + \beta_2 treated + \beta_3 time : treated + \epsilon. \tag{15.1}$$

In this equation, the *time* and *treated* are indicator attributes that would equal 1 for time 2 and for those receiving the treatment, respectively. The treatment effect of interest (the difference-in-differences) would be the β_3 term. This interaction term would estimate the differential effect for those at time two and received the treatment while controlling for the time and treatment effects. The extent to which this term differs from zero would indicate that the treatment group had a differential effect than what would be expected from the control group.

This modeling framework makes the following assumptions that would need to be confirmed in an applied example:

- Parallel trends: absent treatment, treated/control would have changed similarly.
- No anticipation: outcomes don't respond before treatment adoption.
- No spillovers: control units aren't indirectly treated.
- Stable composition if repeated cross-sections.

15.2 More Complicated DID Designs

Recent developments in DID methodology have made it easier to estimate models that have differential timing of the treatment. In this situation, imagine having the same treatment implemented to different groups at different times. Policy changes within an organization could be one example of differential timing. Perhaps a policy was first implemented inside an organization through team A, then implemented with team B later, and so on until the entire organization has the policy implemented. Goodman-Bacon (2021) discussed this idea and proposed using a weighted average of all possible 2x2 DID that could be constructed.

Within DID, it is possible to include more than two groups. For example, suppose there are two treatment groups with a control group. In this situation, each treatment group is compared to the control group similar to above in the 2x2 DID design. With more than one treatment group, the model used to estimate the average treatment effect would expand with more indicator attributes depicting each treatment, further interactions would be established to estimate the average treatment effect after adjusting for time and group effects.

For those interested in reading more about DID are directed toward Cunningham (2021) and Huntington-Klein (2021). This includes discussion of the differential timing situation. There is also some discussion about how best to estimate the standard errors, which would tend to be underestimated (i.e., too small) when using a traditional linear modeling approach.

15.3 Simulating Data from DID

Using the equation depicted above in the simplest 2x2 DID design, this type of design can be simulated with the following model equation: `outcome ~ 1 + time + treatment + time:treatment`. Here the outcome is said to be a function of the time, an indicator attribute that representing before or after the treatment was implemented, and the treatment assignment, another indicator attribute representing whether the group received the treatment or not.

To anchor this example more concretely, imagine the hours worked for hourly healthcare workers across two states in the US were evaluated before and after a state mandated wage increase. In this case, the time attribute would reflect the before and after the state mandated wage increase and the treatment attribute would reflect the two neighboring states in the US. To create these attributes, the fixed portion of the simulation arguments would look like the following:

```
fixed = list(
    time = list(var_type = 'factor', levels = c(0, 1)),
    treatment = list(var_type = 'factor', levels = c(0, 1))
  )
```

Table 15.1 shows the first few rows of the simulated DD data. These data have the treatment and time attributes simulated that are indicator attributes.

TABLE 15.1 First few rows of the difference-in-difference simulated data.

Int	Time	Treat	Time:Treat	ID	Error	Fixed	Random	Hours
1	0	0	0	1	−0.376	32.0	0	31.62
1	0	1	0	2	−0.562	32.2	0	31.64
1	0	1	0	3	−0.344	32.2	0	31.86
1	1	0	0	4	0.090	32.5	0	32.59
1	0	0	0	5	1.599	32.0	0	33.60
1	1	0	0	6	−0.089	32.5	0	32.41

This example sets the baseline for the DD data simulation. More complex scenarios can be built off of this. For example, additional fixed effects (or adding a random structure of states) could be added that represent and adjust for differences across groups within the treatment groups. In the amount of time worked, if the treatment was implemented in more than one state and there were multiple control group states, there may be natural differences in the amount of hours worked in different states. These effects should be removed prior to estimating the average treatment effect.

A number of states could be added to the design by adding a clustering attribute or by adding fixed effects for different states. The choice would be influenced by how the states were selected. For example, if 10 states were randomly sampled, including these as a random effect would better match the design. Instead, if there were four states that were selected purposefully, then those would better be represented as fixed effects. Each of these two designs will be shown below as examples.

15.3.1 Adding Additional Effects

First, consider the example where states were selected purposefully. This could be due to certain states implementing the program of interest or some of the states could have implemented the program while others did not. Since the states were selected purposefully, they would best be represented as fixed effects as they were not randomly selected. These could be simulated with the **simglm** package with the following model equation: `formula = hours ~ 1 + state + time + treatment + time:treatment`. The model equation would estimate differences across states, but the assumption would be that the treatment effect would be the same across states. An interaction could be added that would estimate the treatment difference across the states.

The fixed portion of the model would then look as follows:

```
fixed = list(
    time = list(var_type = 'factor', levels = c(0, 1)),
    treatment = list(var_type = 'factor', levels = c(0, 1)),
    state = list(var_type = 'factor',
                 levels = c('Minnesota', 'Illinois', 'Florida'))
  )
```

In this example, there were three states selected to be included in the study. When simulating the data, not shown here, the default behavior would have two indicator attributes that represent deviations from the reference group. As such, two regression weights would need to be added when simulating the outcome of the attribute. Additional regression weights would need to be added if it is of interest to simulate an interaction between the state effect and the treatment, time, or both.

Instead of selecting states purposefully, suppose that states were selected randomly. More specifically, suppose that fifteen states were selected randomly to collect the data for the study. Instead of including these states as fixed effects, including these as random effects would better match the design. The model formula for this type of specification would look like: `formula = hours ~ 1 + time + treatment + time:treatment + (1 | state)`. This uses syntax that from the **lme4** package for fitting mixed models. The notation in words would specify a random intercept for each state and the variance between states would be estimated.

The random portion of this code would be specified with the following code chunk. The code below specifies the level two random intercept for each state where there is a between state variance equal to 0.3.

```
randomeffect = list(
  int_state = list(variance = 0.3, var_level = 2)
)
```

To finish this design, the sample sizes for the states would need to be specified using the clustered syntax. For this example, the level one sample size would represent the number of observations within each state and the level two sample size would represent the number of states. Often, the sample size within each state would not be equal. This can be included in the simulated data as follows. The sample sizes would use the uniform distribution to sample 15 values that would range between 150 and 400. The `round()` function is used to ensure that these are whole integers. These 15 values would be the sample sizes for each of the 15 states.

```
sample_size = list(level1 = round(runif(15, min = 150, max = 400), 0),
                   level2 = 15)
```

Putting this all together for the randomly selected states would simulate data that looks like the following. The simulated data structure are not shown, but would reflect a structure that is similar to Table 15.1. The primary difference would be an additional columns that represent the random effects associated with the states.

15.4 Estimating Power from DID

The study by Miller et al. (2021) will be used as the primary example for this chapter. This study used survey and administrative data to explore the link between mortality and Medicaid enrollment. The treatment was related to whether states in the United States increased the eligibility of Medicaid as part of the Affordable Care Act. In total, 29 states and the District of Columbia expanded eligibility while the remaining 21 states did not. The states that choose to increase eligibility would serve as the treatment and those that did not would serve as the control. The study focused on individuals aged between 55 and 64 in the year 2014. Ultimately, the study found that the increased Medicaid eligibility reduced mortality rates of the group studied.

To simulate the data to estimate power for this DID effect, Tables 1 and 2 in Miller et al. (2021) show the estimated effect sizes for a variety of different outcomes. Table 1 provides higher level effect size estimates of the treatment effect and Table 2 focuses on the impact from different causes of mortality. For this example, the estimated power will be used to estimate the reduction in mortality that the study found. The DID model estimated by Miller et al. (2021) was:

$$Died = \beta_0 + \beta_1 expansion + \beta_2 post + \beta_3 expansion : post + \beta_s + \beta_t + \beta_j \quad (15.2)$$

In Equation (15.2), the primary term of interest would be β_3 which would represent the differential mortality rate after the Medicaid expansion for the states that expanded eligibility. One reason to power for this term, if the assumed interaction effect is indeed not present in the population of interest, power for the main effect would likely be well powered as well. It would still be a good idea to run the simulation analysis for this as well, but in general, powering a study for an interaction effect will lead to good power for the main effect, too.

There were additional terms that were adjusted for, including the average mortality rate of each state (β_s), the average mortality rate for each calendar year (β_t), and the average mortality rate of each survey wave (β_j). Pulling out the estimate from Table 1 in the primary study, the pooled mortality rate decrease was about 0.132 percentage points for those that expanded compared to those that did not expand eligibility. The study used a general linear probability model and also computed sensitivity estimates from logistic and a cox proportional hazard models. Average mortality rates using the general probability model will be used here to coincide with the primary results from the study shown in Table 1.

The following simulation conditions will be explored to estimate power for the average mortality rate.

- Effect size: 0.1, 0.15, 0.2 percentage point reduction.
- sample size: 2500, 5000, 10000
- Proportion of observations in the treatment: 0.5, 0.4, 0.3

The simulated model will follow the structured model that was fitted to the data, but will assume the data are coming from a single wave and year, which

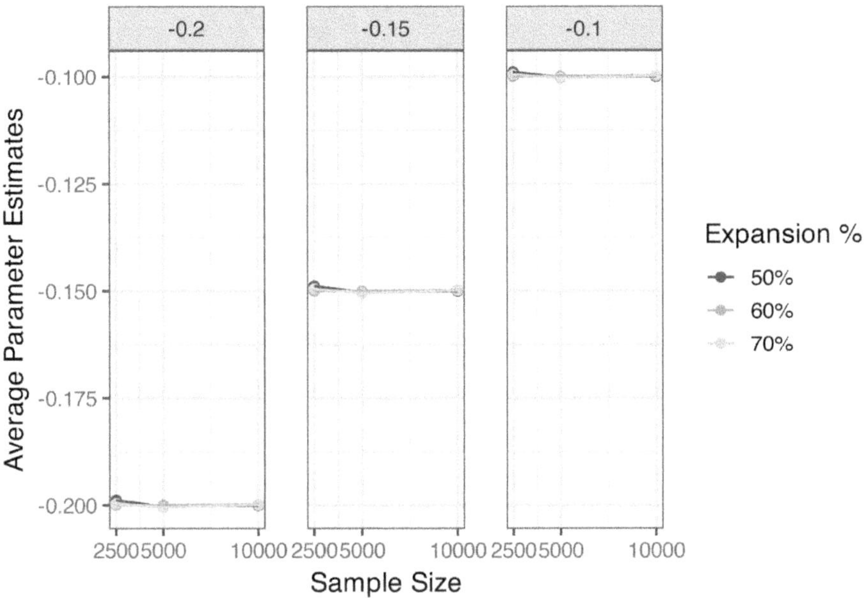

FIGURE 15.1 Average simulated model estimates by sample size, proportion in expansion group, and regression weights.

will remove the β_j and β_t terms from Equation (15.2). In addition, power will be estimated for an average state rather than adjust for state level effects. This would be a great exercise for the reader to develop a new model that also incorporates state effects. In this chapter, mortality_rate is treated as a continuous rate (e.g., deaths per 1,000 individuals), so a linear model is appropriate.

First, exploring the average parameter estimates to make sure that the models are specified correctly and properly recovering the effect size of interest is a great way to start. Figure 15.1 shows the average estimates across the different simulation conditions. The panels in the figure represent the specified effect size for this interaction effect. Overall, the average parameters estimates from the simulated data are close to the specified values, therefore, there are no major concerns about proceeding with the simulated power analysis.

Second, exploring the MSES for the primary term of interest, the interaction effect. Figure 15.2 shows the MSES estimates for the simulated data. As expected, MSES is not strong for any condition with the smallest plausible effect size given the effect sizes thresholds specified. For the -0.15 effect size, the MSES estimates are strong across the larger two sample size conditions for the smallest threshold. For the -0.20 effect size, the estimated MSES power

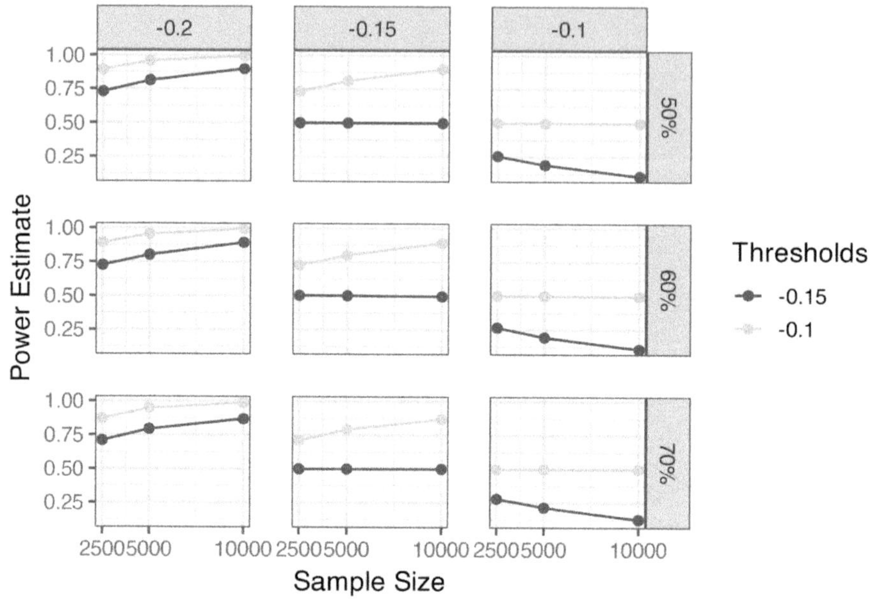

FIGURE 15.2 MSES estimates using the two specified thresholds for practically useful effect sizes by sample size, proportion in expansion group (right panels), and regression weights (top panels).

is strong for all conditions explored. Overall, the percentage in each group doesn't have a large impact on the MSES statistics, likely due to sample sizes being robust regardless of the percentage in the expanded eligibility group. Before making a sample size recommendation, exploring the other statistics would be of interest.

The USE statistics are shown in Figure 15.3. Overall, these are small for most conditions, suggesting the simulated model conditions are able to estimate with strong certainty the direction of the effect of interest. The one exception is the smallest sample size for the smallest specified effect size. Overall, the percentage in each group does not have a strong impact on the estimates.

If this was my study to plan for, I would rule out the smallest sample size and recommend the middle sample size as a trade-off between size and estimated power. This would also allow for some room for missing data to creep in without being too concerned. However, it would be a good idea to further explore sample sizes between 3,000 and 5,000 in more detail to understand potential impacts.

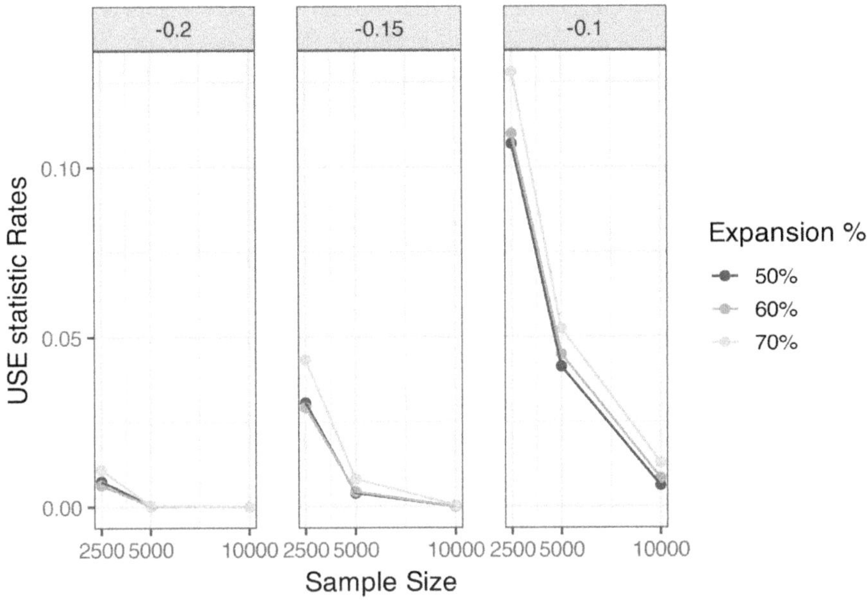

FIGURE 15.3 USE statistic estimates by sample size, proportion in expansion group, and regression weights (top panels).

15.4.1 Robust Standard Errors

It is common in DID studies to use cluster-robust standard errors to reflect the treatment effect often belongs to intact groups at a cluster level. Cluster-robust standard errors address correlation within clusters; they do not, by themselves, fix confounding or violations of parallel trends. In DID, cluster at the level where treatment varies (e.g., state for Medicaid expansion). This is most relevant when your decision rule is based on a p-value (traditional power), but the mechanics are shown here for completeness.

The **simglm** package does allow for the estimation of robust standard errors within the workflow. Below is an example that is run on this with the same simulation parameters, except that random effects are added. The random effects here represent baseline differences across the 50 states. Here we're focusing on robust inference under clustering; adding state FE is a natural extension. I've included the updated "model_fit" simulation arguments below that allow for the estimating of robust standard errors.

To estimate robust standard errors, the following two lines are added to the simulation arguments:

- `robust = TRUE`
- `robust_args = list(cluster = 'state')`

The first argument is a flag to identify that robust standard errors are desired. The second argument is a more general argument that passes more arguments to the **sandwich** package. Ultimately, the **lmtest** package is used to generate the output that is parsed to return estimates for power.

```
model_fit = list(
  formula = mortality_rate ~ 1 + expansion + post + expansion:post,
  model_function = lm,
  robust = TRUE,
  robust_args = list(cluster = 'state')
)
```

The MSES statistics are shown in Figure 15.4. For the two largest population effect sizes specified, the estimated MSES is strong in all conditions except for the largest threshold, where only about 50% of the estimates exceeded this threshold. This isn't surprising given that the population effect size is the same value as the largest threshold. The smallest population effect size has smaller MSES statistics due to them being equal to or smaller than the

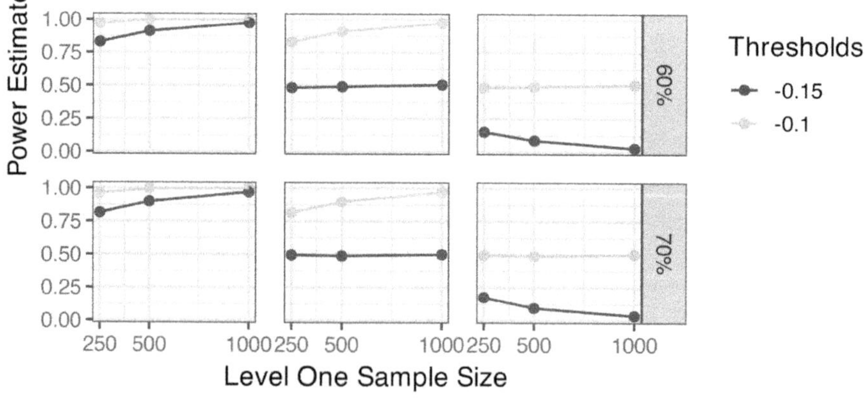

FIGURE 15.4 MSES estimates using the two specified thresholds for practically useful effect sizes by level one sample size, proportion in expansion group (right panels), and regression weights (top panels).

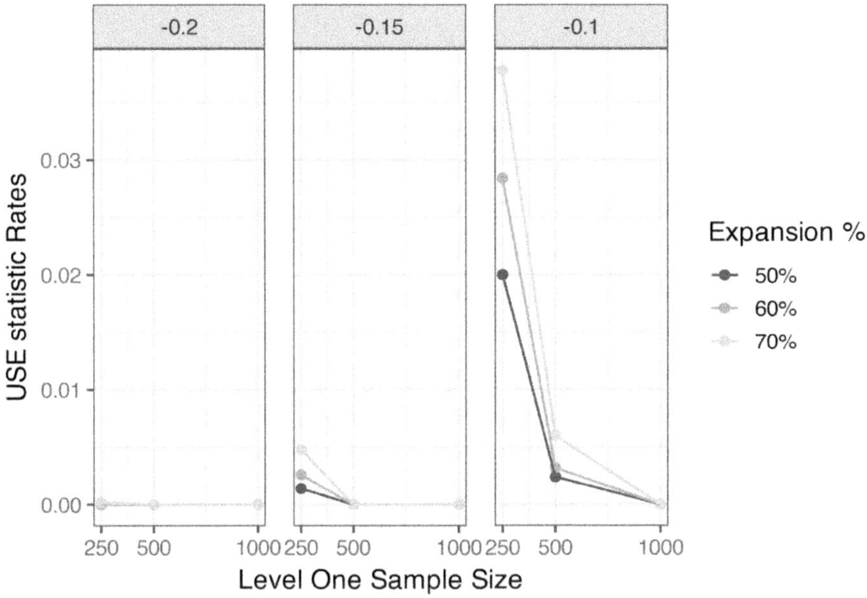

FIGURE 15.5 USE statistic estimates by level one sample size, proportion in expansion group, and regression weights.

practically relevant thresholds defined. The USE statistics are shown in Figure 15.5. Overall, these are strong and would not be of major concern, all occurring less than 3% or so.

16

Propensity Scores

16.1 Chapter Overview

This chapter introduces simulation-based evaluation of propensity score methods for causal inference with observational data. The chapter discusses treatment assignment mechanisms, covariate balance, and common propensity score techniques. Simulation examples illustrate how design choices and model misspecification affect bias and variance. The chapter emphasizes the role of simulation in understanding when propensity score methods perform well and when they may fail.

Propensity scores are another methodology for estimating causal inferences from observational data, sometimes referred to as quasi-experimental designs. Propensity scores attempt to balance groups on observable characteristics with the goal of reducing selection bias concerns. Often propensity scores are used when random assignment was not used or was not ethical to use and selection bias is a concern within the groups. For example, when studying the impacts smoking may have on individual health, it would not be ethical to randomly assign individuals to smoke or not smoke. Any analysis based on these groups would need to use the intact groups already self-selected into those that smoke compared to those that do not. Selection bias would be a concern whenever self-selection occurs rather than random assignment in a true experiment. Propensity scores are one way that researchers can attempt to make groups more equivalent given any observable or measured characteristics of the individuals. Researchers would then still need to articulate the extent to which unmeasured confounders may still be present with respect to selection bias.

The method is implemented by building a model that predicts group membership from observable characteristics. The model is often a logistic regression model when there are two groups, but can be any classification model that can estimate the probability of being in the primary group of interest. After estimating the model, model-implied probabilities of group membership are used to help balance the groups based on the likelihood of being in the group. There are a variety of ways to use the model-implied probabilities to balance the groups based on observable characteristics. These include:

DOI: 10.1201/9781003177876-16

- include the model-implied probabilities in the subsequent model to estimate the treatment effect,
- use the model-implied probabilities as weights in the subsequent models to estimate the treatment effect,
- perform matching using the model-implied probabilities to select observations for each group that are similar,
- stratify the sample based on the model-implied probabilities to reduce variation of units.

For example, imagine that we want to estimate the effect of a treatment to improve road safety. It would be possible to randomly select roads to receive the treatment, however, it would not be possible to randomly have drivers use the roads. Those that live in the area selected to include a new road safety treatment would be more likely to drive on those roads. We could not rule out other factors to ensure comparisons are made on similar driver skill levels. Furthermore, other environmental factors could differ between roads receiving the new safety feature. All these elements could bias the results, motivating adjustment for measured differences before estimating the treatment effect.

Propensity scores could be a way to adjust for sample characteristics if we have data on key external forces that we would wish to adjust for. For example, if there is data on traffic volume, types of traffic, weather factors, and others, we could estimate a propensity score model prior to estimating the treatment effect.

The propensity score analysis is not limited to two groups, the same idea can be generalized to more than 2 groups. For example, if there were two treatment conditions and a control group. In this situation, the propensity score model used would still predict group membership, but would need to handle more than two groups. Instead of a logistic regression, a multinomial logistic regression could be used. A classification model, such as random forest, could still be used, but the model would predict more than two groups.

For more information on the propensity score methodology, see these resources Austin (2011), Li et al. (2018), Rosenbaum and Rubin (1983), and Stuart (2010).

16.2 Simulating Data from Propensity Scores

The **simglm** package introduces a few new simulation arguments to aid in the two-step process to simulate data from a propensity score design. One key element of this approach is that any data attributes simulated as part of the simulation process are carried over to the primary data generating model. Much of the other mechanics used internally are the same as already discussed, but the simulation arguments are specified within a new named element.

The simulation arguments for propensity scores are specified in a named list element, `propensity`. This element of the simulation arguments takes all possible elements discussed throughout the simulation section of this book. Let's look at a single level example to start building out this process.

The following example is a full example that generates the data for a propensity score model. Note, this does not generate a full population model, rather, it just does the propensity score simulation procedure. In the next section we will combine this to include the second portion in addition to the propensity score simulation. Notice that the named argument of the list, "propensity," is used here. Within this list element, all previous simulation arguments can be specified. The ones that are required include:

- **formula:** The propensity score model formula,
- **fixed:** The fixed effect portion of the propensity score model, defining the key data attributes on the right-side of the model formula,
- **sample_size:** required unless inherited from the main simulation arguments,
- **reg_weights:** The regression weights that relate the fixed effect data attributes to the outcome,
- **outcome_type:** The outcome type, for binary treatments, this would be 'binary', which would differ from the continuous default.

In the following simulation arguments, the error argument is optional. If this is not specified, additional error is not added when the outcome type is specified as binary. The error simulation argument can add some additional uncertainty to the propensity score model to reflect data that is not collected but impacts the treatment effect designation. More explicitly, this would not represent measurement error in the treatment allocation, instead it would represent unmeasured attributes that are also influencing treatment assignment. In this case, this error is random, meaning that it will not, by itself, systematically bias the estimated treatment effect via outcome-related confounding (i.e., induce imbalance on outcome-relevant covariates), but it will make the propensity simulation model be less deterministic and in my view, more reflective of real-world data conditions. Including the error variance will typically increase the amount of overlap between the treated and control groups and will increase the variance of the estimated treatment effects.

```
sim_arguments <- list(
    propensity = list(
        formula = trt ~ 1 + age + ses,
        fixed = list(age = list(var_type = 'ordinal',
                                levels = -7:7),
```

```
                                ses = list(var_type = 'continuous',
                                           mean = 0, sd = 5)),
        sample_size = 15,
        error = list(variance = 5),
        reg_weights = c(2, 0.3, 0.15),
        outcome_type = 'binary'
    )
)

simulate_propensity(sim_arguments) |>
  head()
```

```
##   X.Intercept. age    ses level1_id   error
## 1            1   7  6.403         1 -2.4277
## 2            1   7 -8.636         2 -0.1910
## 3            1  -5  8.451         3  2.3940
## 4            1   6  2.519         4 -0.3251
## 5            1  -5 12.642         5 -2.6062
## 6            1   2  2.745         6 -1.8303
##   fixed_outcome random_effects untransformed_outcome
## 1         5.060              0                 2.633
## 2         2.805              0                 2.614
## 3         1.768              0                 4.162
## 4         4.178              0                 3.853
## 5         2.396              0                -0.210
## 6         3.012              0                 1.182
##   trt
## 1   1
## 2   1
## 3   1
## 4   1
## 5   0
## 6   1
```

The resulting output should look similar to what has already been shown earlier. Key data columns representing the simulation model are specified to be as transparent as possible. When this is built into a larger propensity score simulation pipeline, these data will be held constant and not resimulated, a key element to make sure the data do not shift across the two steps.

16.2.1 Building the Full Simulation Pipeline

The rest of the simulation pipeline builds off of all the elements that have already been discussed throughout the book. All of the other simulation arguments are possible to be used here, the key difference from other generating processes is the inclusion of the `propensity` simulation arguments. The propensity arguments are processed before the other simulation arguments when present and anything simulated as part of the propensity score model are not regenerated.

One way to see this in the following code is within the fixed effect argument. In the fixed effect simulation argument within the propensity section, the age and ses are specified. However, the treatment, age, and ses are not specified in the fixed effect simulation argument outside of the propensity score arguments. This would typically result in an error, but in this case, since the propensity score arguments are simulated first, they are not needed to be re-simulated and are accessible to the broader simulation pipeline. Lastly, the sample size argument does not need to be specified directly within the propensity score arguments. Instead, if it is omitted, the sample size is automatically filled in from the general simulation arguments. This is especially useful as a larger simulation is designed that includes sample size as a varying argument.

```
sim_arguments <- list(
    formula = achievement ~ 1 + motivation + trt + age + ses,
    fixed = list(
        motivation = list(var_type = 'continuous',
                          mean = 0, sd = 20)
    ),
    sample_size = 15,
    error = list(variance = 10),
    reg_weights = c(50, 0.4, 1.2, 0.1, 0.25),
    propensity = list(
        formula = trt ~ 1 + age + ses,
        fixed = list(age = list(var_type = 'ordinal',
                                levels = -7:7),
                     ses = list(var_type = 'continuous',
                                mean = 0, sd = 5)),
        error = list(variance = 5),
        reg_weights = c(2, 0.3, -0.5),
        outcome_type = 'binary'
    )
)
simulate_fixed(data = NULL, sim_args = sim_arguments) |>
    simulate_error(sim_args = sim_arguments) |>
    generate_response(sim_arguments) |>
  head()
```

```
##   X.Intercept. motivation trt age     ses level1_id
## 1            1    -9.7785   0  -2  6.8430         1
## 2            1   -12.5590   1  -6 -1.1289         2
## 3            1    -0.9383   0  -7  7.5824         3
## 4            1     3.2524   0  -6 -7.7438         4
## 5            1    25.8461   1  -4  2.9231         5
## 6            1    -9.2711   1   5  0.6193         6
##     error fixed_outcome random_effects achievement
## 1 -2.1035         47.60              0       45.50
## 2  1.4294         45.29              0       46.72
## 3  1.6661         50.82              0       52.49
## 4 -0.7282         48.77              0       48.04
## 5  4.4191         61.87              0       66.29
## 6  5.5772         48.15              0       53.72
```

Notice in the output that the primary outcome, achievement is generated and includes all the other data elements associated with them. The data attributes, trt, age, and ses, were simulated as part of the propensity score simulation. Then, these were again used to simulate the population model to generate the final achievement outcome score.

16.2.1.1 Showing the Treatment Dependency

It is important to consider and explore how the assignment of the treatment attribute was generated from the propensity score model. This was step one of the simulation procedure to generate the treatment allocation via a model-based assignment mechanism rather than through random assignment. For example, in the above example, treatment assignment was driven by the age and SES of the individual. Figure 16.1 shows that those in the treatment group, values of 1 in the figure, were from lower SES backgrounds compared to those in the control group. This was controlled by the negative coefficient in the propensity simulation arguments. Practically, this could reflect a situation where those who receive the treatment are from lower-income backgrounds compared to those who did not receive the treatment. Statistically, this means that the SES background differs and thus there is not sufficient balance for those who received and did not receive the treatment. If the analysis were to proceed without considering this, the treatment effect could be biased and we could further not rule out SES differences as a potential cause of differences in the outcome. This figure does not show the estimated treatment effect, rather it shows the imbalance that propensity scores can be one way to adjust for this imbalance statistically.

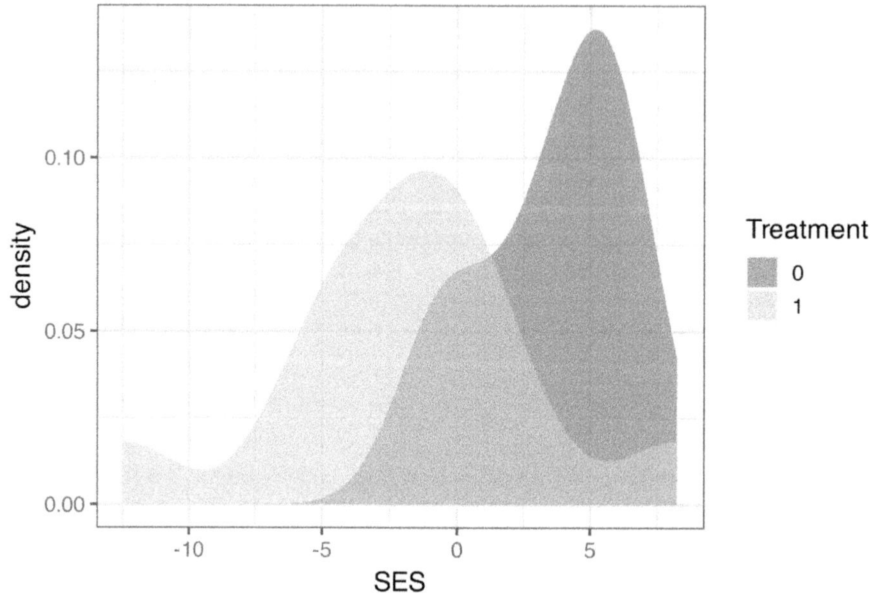

FIGURE 16.1 Density figure showing the non-random assignment of individuals to the treatment group.

```
gf_density(~ ses, data = prop_data, fill = ~ factor(trt)) |>
  gf_labs(x = "SES",
          fill = 'Treatment') |>
  gf_refine(scale_fill_grey())
```

16.2.2 Extending to Multilevel Cases

The **simglm** package allows for multilevel propensity score simulation. The treatment or group assignment can be done at any level of the multilevel structure as well. For example, if the intact groups occur at level one, two, or three, these can be specified through optional simulation arguments within the `propensity` simulation arguments.

Only one more additional simulation argument is needed to simulate a multilevel propensity score model. The `randomeffect` argument specifies the simulation characteristics of the random effects specified. A few others need to be modified to specify the appropriate multilevel structure. As a reminder, this follows the **lme4** package formula specification using parentheses with a vertical bar (see the example on the following page for a reminder). To also specify the multilevel structure, additional sample size arguments are needed to state the number of level one and two units to be simulated.

Below is an extended simulation example that specifies the multilevel propensity simulation arguments. These would be inserted into the broader simulation arguments to pass into the **simglm** simulation pipelines.

```
propensity = list(
        formula = trt ~ 1 + age + ses + (1 | classroom),
        fixed = list(age = list(var_type = 'ordinal',
                          levels = -7:7),
                      ses = list(var_type = 'continuous',
                          mean = 0, sd = 5,
                          var_level = 2)),
        sample_size = list(level1 = level1_samp, level2 = 50),
        error = list(variance = 5),
        randomeffect = list(int_classroom =
                              list(variance = 5, var_level = 2)),
        reg_weights = c(2, 0.3, -0.5),
        outcome_type = 'binary'
    )
```

The default behavior is to specify the group or treatment to the level one units. There are also additional optional arguments that allow the treatment or group membership to be at level two (or three if specified). To specify the treatment membership to be at a different level, the `outcome_level` simulation argument is used. Numeric values are used to specify different level, for example, to specify the treatment to be assigned at level two, `outcome_level = 2` can be added to the list of propensity simulation arguments. If a level three model was specified, the outcome level can be either two or three, depending on what is desired.

The following are lists of examples that the outcome_level simulation argument could be used to align the treatment to.

- Treatment at level 1: student or individual self-selection to attend classes.
- Treatment at level 2: classrooms teachers select the curricula, programs select training.
- Treatment at level 3: school curricula or policy choice, states adopt health policy.

There could be many more structures where this is useful, but the goal of the outcome_level simulation argument is to allow users to adapt the treatment level to align with their specific example. This can enhance the number of situations that the **simglm** package can be used for. Also, typically, treatment allocation at a higher level unit comes with more data needs to achieve a similar level of power.

16.3 Estimating Power from Propensity Scores

Estimating power from propensity scores takes a similar process to simulating propensity scores. Namely, a two step process is used, first a propensity score model is fitted and the propensity scores are extracted. Then, after estimating the propensity score model, the final model is used to estimate the treatment effect.

In order to perform this two stage process, another named simulation argument was created. The `propensity_model` simulation argument is used to signal that a propensity score design and two-stage estimation should be used.

Below is an example `propensity_model` simulation arguments. These arguments specify both the model used to estimate the propensity scores and the adjustment strategy applied in the outcome analysis. In this example, covariate adjustment using the estimated propensity score is shown for simplicity.

Alternatively, researchers may specify weighting-based adjustments, including inverse probability weighting or stabilized inverse probability weighting, by setting:

- 'propensity_type = 'ipw'
- 'propensity_type = 'sbw'

The choice of propensity score adjustment method can influence both covariate balance and the precision of the resulting treatment effect estimates. In particular, the performance of each method depends on how well the propensity score model aligns with the underlying data-generating process. From a study planning perspective, simulation provides a natural framework for evaluating which adjustment strategy performs best under plausible modeling assumptions (Austin, 2011).

```
propensity_model = list(
        formula = trt ~ 1 + age + ses,
        propensity_type = 'covariate'
    )
```

16.3.1 Real-World Example

To give a real-world example of estimating a power analysis for a propensity score approach, we will use the average treatment effects estimated from Sullivan and Field (2013). The treatment is defined using an intent-to-treat framework in which children are classified as treated if they were identified as receiving preschool special education services, regardless of the specific services, duration, or intensity received. This avoids conditioning on post-assignment characteristics and aligns the analysis with the average treatment effect on the treated (ATT), which reflects the effectiveness of typical preschool special education services as implemented in practice. Using a propensity score weighting approach, they estimated ATT effects of −0.21 for reading and −0.28 for mathematics.

To estimate power for this type of study, we need to specify what type of model we are interested in fitting. The study, after estimating the propensity score model and using the propensity score weights, only included the treatment attribute in the final model with the outcomes of interest. The other data attributes were used solely to estimate the propensity score model. To estimate power, I'll do a similar type of design. To reflect realistic rates of special education, we will design the model to be unbalanced with about one special education student to every nine students not receiving special education services.

For terms to vary, these are the conditions varied in this example

- **Effect size:** −0.15, −0.20, −0.25, −0.30, −0.35, −0.40
- **Sample size:** 200, 500, 1000, 2000
- **Propensity error variance:** 1, 3, 5

Only single level regression models will be explored in this example. It is left to the reader to generalize this to a multilevel structure. Ultimately, 72 conditions will be fully crossed to explore, each condition will be replicated 5,000 times. Here is the simulation code to be able to conduct this analysis. The code for the propensity model as well as the varied propensity specifications are shown below. For the propensity model, covariate adjustment was used, however, weighting is a valid approach that would be a good option to explore in tandem with the covariate adjustment. Only the variance equal to 1 is shown for the propensity score model section, but two additional simulation conditions showing additional error variance were specified.

```
propensity_model = list(
  formula = special_education ~ 1 + parent_engagement + ses,
  propensity_type = 'covariate'
)
vary_arguments = list(
  propensity = list(
    list(
      formula = special_education ~ 1 + parent_engagement + ses,
      fixed = list(parent_engagement = list(var_type = 'continuous',
                                             mean = 0, sd = 3),
                   ses = list(var_type = 'continuous',
```

```
                    mean = 0, sd = 5)),
      error = list(variance = 1),
      reg_weights = c(-3, -0.5, -0.3),
      outcome_type = 'binary'
    )
  )
)
```

The estimated parameters were compared to the specified simulation parameters to ensure the models are being simulated and estimated well. Overall, there was strong alignment between the average estimated parameters and the simulation arguments specified. As such, there is no concern about proceeding with the simulated power analysis below.

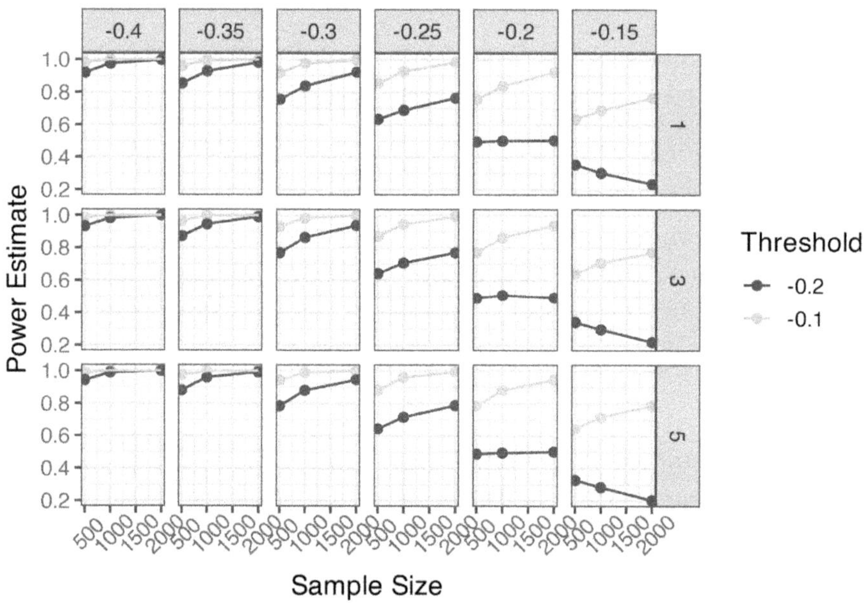

FIGURE 16.2 Alternative power estimates using the two specified thresholds for practically useful effect sizes by sample size, propensity variance, and regression weights.

Figure 16.2 shows the MSES statistics from this simulated propensity score design. The sample sizes are shown on the x-axis, the propensity score error variance is shown on the row panels, and the different regression weights are shown in the column panels. The figure highlights the sample size and power tradeoffs, with increasing power to suggest the effect size is larger than the specified thresholds. The opposite is true for effect sizes that are −0.20 or smaller for the −0.20 threshold. In this instance, statistical power to detect the desired effect decreases due to an increase in precision. More specifically, in the case with a sample size of 2,000, there are fewer estimates that are larger than −0.20 when the true effect size is −0.15 due to the increase in precision. This highlights a case that we wouldn't be interested in and would want to be filtered out to visualize, but is kept here for illustrative purposes.

Overall, for the smallest threshold, −0.10, power estimates are strong for all sample sizes of 2,000, but smaller sample sizes produce less likelihood of differentiating from −0.10 when the effect sizes are small. When the effect sizes are greater than −0.25, all power scenarios show strong estimates. For effect sizes that are less than −0.25, sample sizes close to 2,000 would be needed in many scenarios to have the traditional power threshold of 0.80.

We can also explore the USE statistics, which at a minimum we would want to be very small for any study. Figure 16.3 shows the USE statistics, or the proportion of effects that would be different from the hypothesized population effect. As would be expected, those with smaller effect sizes and smaller sample sizes would have larger USE statistics. The estimates approached 0.15 for a population effect size of −0.15. Practically, this means that about 15% of the simulated effects are positive rather than the negative specified by the population model. As the effect sizes increase, the USE statistics decrease substantially to not be of practical concern for any of these sample sizes.

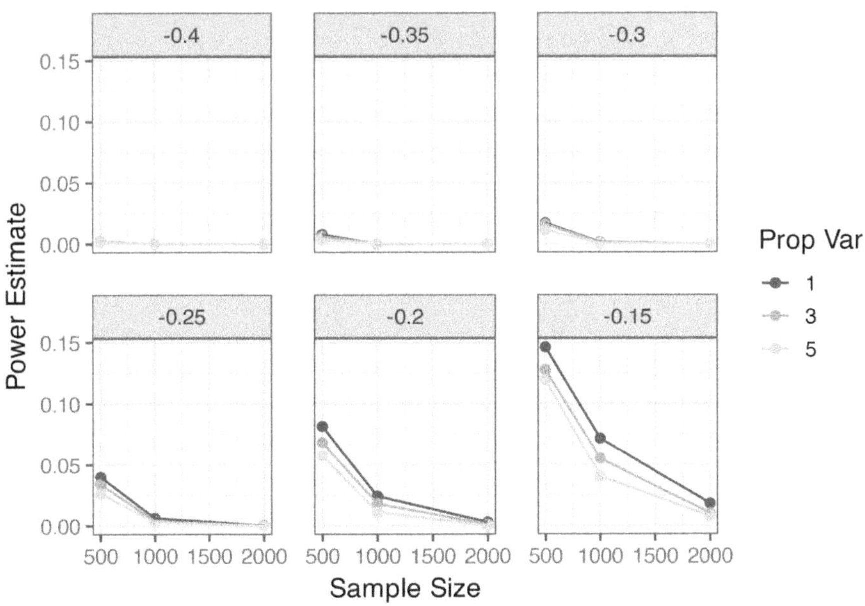

FIGURE 16.3 USE statistic estimates by sample size, propensity variance, and regression weights.

Given the simulation approach, one could also explore the distribution of coefficient estimates across the 5,000 replications for each data condition. This is left to the reader to complete in this instance, but this would further show the USE statistics as well as the alternative power estimates greater than the specified thresholds. Another useful step when exploring the estimating power for this design would be to establish how much differences there are between the covariate adjustment compared to the inverse probability weighting schemes.

17

Non-Experimental Designs

17.1 Chapter Overview

This chapter synthesizes simulation approaches across a range of non-experimental and quasi-experimental designs. The chapter highlights common challenges such as confounding, selection bias, and model dependence. Simulation is presented as a tool for evaluating assumptions, comparing analytic strategies, and understanding the limits of causal inference without randomization. The chapter reinforces the role of simulation as a decision-support tool rather than a guarantee of causal validity.

Non-experimental designs are those in which treatment or exposure is not randomly assigned, and inference relies on statistical modeling rather than design-based control (Cook et al., 2002). These studies are also described as correlational or observational (Cook et al., 2002). The defining feature of non-experimental designs stems from the lack of randomization and, therefore, the lack of strong causal conclusions that can be drawn from standard analyses. As such, these studies are often exploratory and may benefit from a future follow-up study that introduces randomization or uses observational causal inference methods to better isolate the effect of interest.

Even though non-experimental designs are often unable to support strong causal claims, they are common, particularly in the social sciences. In many settings, randomization is not feasible; therefore, non-experimental designs may be the only practical option, sometimes alongside other approaches for causal inference with observational data (e.g., regression discontinuity or propensity scores). In this chapter, power refers to detecting adjusted associations (conditional model parameters), not causal effects. We treat coefficients as describing differences in outcomes conditional on included covariates and model assumptions.

When correlational or non-experimental designs are used, there are often many threats to causal claims, sometimes referred to as internal validity (Cook et al., 2002), that are difficult if not impossible to rule out. The following are a few common threats to internal validity within non-experimental designs:

DOI: 10.1201/9781003177876-17

- confounding and selection bias (systematic differences between exposure/comparison groups, including unmeasured causes of both exposure and outcome)
- regression to the mean, a phenomenon where those with extreme scores on one or more constructs will be less extreme on other measures
- temporal precedence / reverse causation, no clear distinction about which attribute occurred first to determine cause and effect

Without ruling out these threats to internal validity, strong claims cannot be made. However, non-experimental designs can be important for exploratory research to understand if there are associations that should be studied in a more controlled fashion.

When simulating data from a non-experimental design, all of the strategies discussed earlier in the book would be valid to be used. The primary differentiator is that there is no random assignment, therefore, specific care in how the treatment groups are generated should be made.

17.2 Estimating Power for Non-Experimental Designs

To estimate power for non-experimental designs, a study by LeBeau et al. (2012) will be used as an example. This study explored student characteristics that were associated with graduating college with a science, technology, engineering, or mathematics degree. The study used a logistic mixed model to explore how the likelihood was impacted by the student and high school characteristics. A few of the key characteristics that this study explored were standardized test scores (ACT mathematics score), high-school mathematics performance, and demographic attributes at the student level as well as high school mathematics curriculum and high school location (i.e., rural, suburban, or urban settings).

This example will depict a binary outcome, graduated with a STEM degree or not, and will have focal attributes of interest at level one (student level) and level two (high school level). The simulating model explored initially below assumes that these terms are the only ones specified and that there are no interactions between any of the model terms. Unobserved heterogeneity was specified (`variance = 5`; specified via error simulation argument) to represent omitted predictors or other unmodeled influences on the outcome. Although logistic models do not include an additive residual variance parameter in estimation, simulation can incorporate unobserved heterogeneity by adding a normally distributed disturbance to the linear predictor prior to applying the logistic link. This does not change the fitted model; it makes the data-generating process more realistic by introducing omitted-variable variation.

Furthermore, the nested effect of high schools on individuals was specified with a variance of 3 for the random intercept. Notice that the number of replications was specified as 5,000. It is recommended that this should be the minimum number of replications specified as increasing the number of replications would increase precision.

```
set.seed(123)

level1_samp <- round(runif(200, min = 5, max = 35), 0)

sim_arguments <- list(
  formula = stem_degree ~ 1 + act + male + math_curriculum +
    (1 | high_school),
  fixed = list(act = list(var_type = 'ordinal', levels = 14:36),
               male = list(var_type = 'factor',
                           levels = c('female', 'male')),
               math_curriculum = list(var_type = 'factor',
                                levels = c('NSF', 'UCSMP', 'CD'),
                                var_level = 2)),
  randomeffect = list(
    int_highschool = list(variance = 3, var_level = 2)
  ),
  sample_size = list(level1 = level1_samp, level2 = 200),
  reg_weights = c(-6.308, 0.12, 0.5, -0.02, -0.26),
  error = list(variance = 5),
  outcome_type = 'binary',
  replications = 5000,
  model_fit = list(
    model_function = lme4::glmer,
    formula = stem_degree ~ 1 + act + male + math_curriculum +
      (1 | high_school),
    family = binomial),
  extract_coefficients = TRUE,
  power = list(
    dist = 'qnorm',
    alpha = 0.01,
    direction = 'two-sided'
  )
)

replicate_simulation(sim_arguments) |>
  compute_statistics(sim_arguments, type_1_error = FALSE)
```

TABLE 17.1 Power estimation for non-experimental design with 5,000 replications.

Term	Avg. Estimate	Est. Power	Repl
(Intercept)	−3.86	1.00	5000
act	0.07	1.00	5000
malemale	0.31	0.81	5000
math_curriculumCD	−0.17	0.09	5000
math_curriculumUCSMP	−0.01	0.07	5000

Table 17.1 shows the power results for this model using the NHST framework. Four of the total output columns are shown representing the term, the average parameter estimate across the replications, the estimated power, and the number of replications. In this situation, the terms that would be of most interest would be associated with the math curriculum and would represent the association that the CD and UCSMP curriculum would have on the likelihood an individual would graduate with a STEM degree. The average estimates are both negative, suggesting that the likelihood of individuals graduating with a STEM degree and had those two curricula was smaller compared to the NSF curriculum. As the goal is to perform study planning here, given the simulated data, the estimated power would be about 0.09 and 0.07 for the CD and UCSMP model estimates being different from 0 at the 0.01 p-value threshold. In contrast, if we were also interested in evaluating power for the male or ACT terms, the estimated power for these two terms were greater than 0.80, an often used threshold for determining sufficient power.

Not shown here as output, if α was specified as 0.05 instead of 0.01, then the estimated power for the math curriculum attributes increases to 0.17 and 0.11 for the CD and UCSMP curricula being different from the NSF curriculum. The other two terms, the sex or ACT terms, power increases to above 0.90 for the sex term. In non-experimental designs, small effects plus clustering often imply that power is primarily driven by the number of level-2 units and covariate quality, not just total sample size.

17.2.1 Alternative Power Estimation

As discussed in Chapter 10, power estimation by simulation gives much more flexibility with respect to how the power estimation is computed. For this example, it may be of interest to see how many of the parameter estimates are large enough to be of practical interest. These may vary for different people, but parameter estimate threshold(s) can be specified in order to compute the proportion of the parameter estimates that are at least as large as the threshold(s) specified.

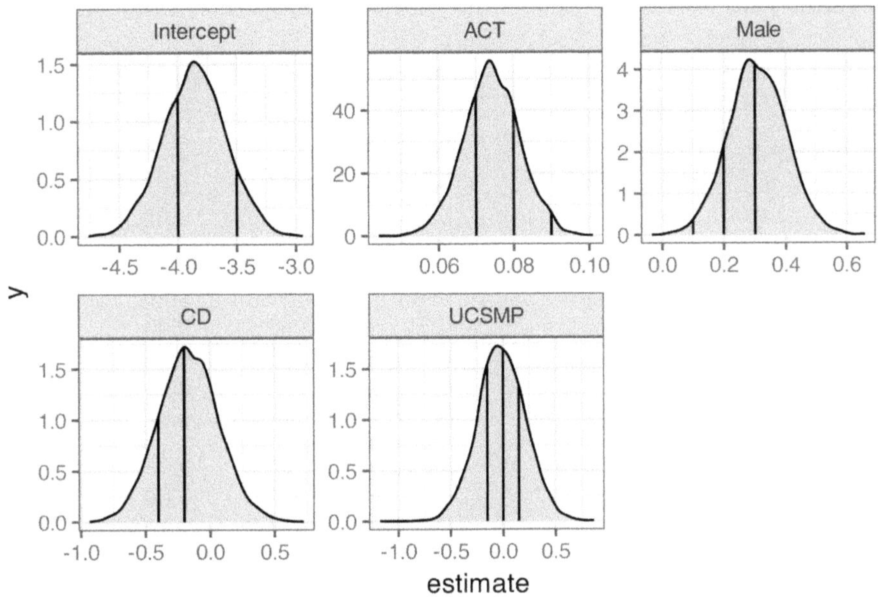

FIGURE 17.1 Density plots of simulated parameter estimates for estimating power.

Figure 17.1 shows the alternative power estimation. Values are specified that would represent different magnitude of effects that are desired. The vertical lines within the density figure would represent the various thresholds that are specified as being values of interest for effects that would be practically meaningful. Focusing on the terms associated with the math curriculum, notice how the distribution for the CD term has a wide range of estimated values in the simulated distribution. The thresholds specified, -0.4 and -0.2 indicate that less than half of the distribution are outside of these thresholds indicating large or medium effects, respectively. This would suggest that more than half of the time the effect would be estimated as small or negligible or positive.

When exploring the UCSMP term, notice that the distribution of estimated effects is centered closer to zero. This indicates that about half of the time the effect is estimated as positive and the other half negative. The effect size specified was -0.02, therefore, it makes sense that this term would be centered close to 0. Less than a third of the distribution is shown to be medium effect size of -0.15 or 0.15, respectively.

When exploring the distribution of the sex and ACT terms show evidence of much larger effects. For example, when exploring the estimated ACT effects, if an association of 0.07 was deemed a small effect size, then about 70% of the distribution would be estimated in this range. If 0.09 was deemed a large effect

TABLE 17.2 Alternative power estimates based on specified threshold values.

term	avg_estimate	alt_power	threshold	replications
(Intercept)	−3.86	0.30	−4.00	5000
(Intercept)	−3.86	0.91	−3.50	5000
act	0.07	0.72	0.07	5000
act	0.07	0.21	0.08	5000
act	0.07	0.02	0.09	5000
malemale	0.31	0.99	0.10	5000
malemale	0.31	0.88	0.20	5000
malemale	0.31	0.53	0.30	5000
math_curriculumCD	−0.17	0.16	−0.40	5000
math_curriculumCD	−0.17	0.45	−0.20	5000
math_curriculumUCSMP	−0.01	0.27	−0.15	5000
math_curriculumUCSMP	−0.01	0.47	0.00	5000
math_curriculumUCSMP	−0.01	0.24	0.15	5000

size, then less than 2% of the estimated distribution would be estimated as large. Similar interpretations would be made the sex term (labeled as Male), except in this case a majority of the estimated distribution would reflected as a small effect (about 99%), about 88% as a medium effect, and over half of the distribution was estimated as a large effect.

Finally, when thinking about USE statistics in this context, the distributions show very little or no evidence of USE statistics. Only the sex effect shows an effect size of zero in the estimated distribution. However, USE statistics would be much more common for the math curriculum effect sizes. USE statistics would make up about half of the UCSMP distribution, whereas the CD distribution shows evidence of 30 to 40% of the distribution making a USE statistic.

Table 17.2 shows the MSES statistics. These estimates reflect the proportion of the distribution equal to or greater (or less for negative thresholds) than the thresholds specified in the power arguments.

Table 17.3 shows the estimated USE statistics for each term in the estimated model. Recall, the direction of the effect sizes are specified within the power simulation arguments. As discussed above, the intercept, ACT, and sex terms were all practically zero, indicating that USE statistics (change in estimated sign of effect size) do not occur. USE statistics do occur more frequently for the math curriculum terms, suggesting that the estimated effect is not as clear as for the other terms. Given these are the primary estimated effects and assuming that these are expected to not be equal to zero, having USE statistics that

TABLE 17.3 USE statistics based on specified direction of effect sizes.

term	avg_estimate	type_use	sign	replications
(Intercept)	−3.86	0.00	negative	5000
act	0.07	0.00	positive	5000
malemale	0.31	0.00	positive	5000
math_curriculumCD	−0.17	0.23	negative	5000
math_curriculumUCSMP	−0.01	0.47	negative	5000

approach 50% in the case with the UCSMP math curriculum would suggest that some additional study planning may need to occur to increase the sample size or re-evaluate the magnitude of the effect size.

17.3 Varying Simulation Arguments

When conducting a proper power analysis, it is often of interest to vary different parameters to understand their impact on the power estimates. Often, sample size is one simulation argument that is varied to determine the extent to which that impacts power estimates and if the planned sample size is large enough to find the desired effect.

Within the non-experimental study design, other elements would likely want to be varied to understand the impact they have on power estimation. For example, the size of the regression coefficient or effect size would be one that may be fruitful. For example, since the design is non-experimental, the effect may be biased, therefore, care needs to be taken to explore the impact the power estimates may have if the effect size is smaller. Furthermore, getting a good estimate for the unobserved heterogeneity variance is also crucial to explore to understand how much this could impact power estimates. Often, it would be of interest to explore the impact larger error variances may have on the power estimates as this would increase the standard error for the parameter estimates, which would reduce parameter estimate precision.

Other conditions that could be useful to vary include:

- the amount of missing data,
- population generating model,
- sample fitted model,
- measurement error,
- omitted confounder strength,
- cluster imbalance,
- model misspecification.

The more things that are varied, the more likely the power estimates would be as close to the sample drawn and be as accurate as possible. The examples from the previous example can be built into these examples and are left as an exercise for the reader.

18

Simulation as a Framework for Study Design and Causal Reasoning

18.1 Chapter Overview

This concluding chapter revisits simulation as a unifying framework for study design and methodological reasoning. The chapter synthesizes themes from throughout the book, emphasizing how simulation supports transparency, robustness, and informed decision-making. The discussion highlights how simulation complements causal inference frameworks and encourages researchers to move beyond rigid analytic formulas. The chapter concludes by positioning simulation as a practical, extensible approach for applied researchers navigating complex design and analysis choices.

This book has used simulation as a framework for moving beyond basic calculations for study design to a more thorough investigation of tradeoffs. Statistical analysis is more than a collection of recipe cards tailored to specific problems; it is a sequence of interconnected decisions and trade-offs that shape the final analysis. Simulation as a tool can mirror these decisions to understand their impact on the full study design phase. Coupling simulation with causal inference and experiments further enhances the types of decisions that can be made and further increases the strength of the studies that are currently being explored.

18.2 Simulation Beyond Power

Simulation has been a tool to estimate power in the past. Prior work in this area has focused on using simulation as a tool to decide what proportion of hypothesis tests are rejected. This sells the benefits of simulation short. Instead, this book takes a different approach, anchoring the exploration of simulation and the benefits of it within a framework that utilizes it as a way to enhance study planning.

DOI: 10.1201/9781003177876-18

Simulation provides a more powerful framework to quantitatively explore how different statistical decisions, model assumptions, or other tradeoffs have on detecting the effect of interest. Not only does this align more closely with how a strong power analysis should be explored, it also more explicitly maps the study design, assumptions, causal pathways, and models fitted to the research questions being answered.

The difficulty of adopting a simulation-based approach is also its strength. It transcends the 'plug-and-play' nature of traditional power calculators, which often encourage a 'black-box' approach to design. Instead, it requires researchers to be explicit about the data generating process, the study design features, the population they wish to study, and how likely the data to be collected are to satisfy the underlying statistical assumptions. The level of explicitness that is needed here is the strength that increases the transparency and thoroughness of the analysis, leading to stronger understanding of tradeoffs and their impact on the estimated power.

Table 18.1 summarizes key differences between traditional and simulation-based power analyses as they are framed in this book.

TABLE 18.1 Comparison of traditional and simulation-based power analysis.

Feature	Traditional Power Analysis	Simulation-Based Power Analysis
Primary goal	Determine sample size for a prespecified test	Evaluate design trade-offs and inferential properties
Underlying model	Typically closed-form or asymptotic	Explicit data-generating process
Model complexity	Limited to standard analytic forms	Flexible; constrained by model estimability
Assumptions	Often implicit or fixed	Explicit and directly encoded
Design features	Difficult to incorporate beyond simple settings	Naturally accommodates clustering, longitudinal structure, and heterogeneity
Missing data	Typically handled indirectly or ignored	Can be modeled explicitly
Sensitivity analysis	Limited or ad hoc	Central to the analysis
Interpretation	Focused on rejection probabilities	Focused on robustness and uncertainty
Computational burden	Low	Moderate to high

18.3 What Simulation Can and Cannot Do

Even though simulation is a powerful aid in study design, it cannot do everything. Simulation is not a replacement for subject-matter expertise, does not guarantee external validity, and does not validate identification. Because simulation requires the researcher to specify a data-generating process, it necessarily operates within a closed world defined by modeling assumptions. Identification assumptions must therefore be imposed rather than discovered, making simulation unsuitable for validating causal identification. The plausibility of simulated scenarios depends critically on subject-matter expertise, which cannot be replaced by computational sophistication. Moreover, robustness of results within a simulated setting does not ensure external validity across populations, contexts, or time. For these reasons, simulation is best viewed as a tool for structured reasoning about design trade-offs rather than a guarantee of causal validity.

18.4 Implications for Study Planning and Applied Research

Viewed as a framework rather than a procedure, simulation encourages researchers to approach study planning as a sequence of interconnected decisions rather than a single calculation. Questions about sample size, model specification, statistical assumptions, and statistical models are not independent, and simulation provides a way to explore how these choices interact under plausible simulation arguments.

In practice, this means that simulation is most valuable when key features of a study or population are uncertain or contested. This could be when planning a study on a new treatment or exploring the impact of a treatment on a new population. When identification assumptions are strong and the design closely matches standard analytic settings, simpler power calculations may be sufficient. However, as designs become more complex, through clustering, longitudinal structure, treatment heterogeneity, or reliance on quasi-experimental assumptions, the benefits of simulation increase substantially. Closed-form solutions often rely on assumptions that are either inherently under scrutiny or unlikely to hold in practice (e.g., sphericity).

Importantly, the goal of simulation-based study planning is not to identify a single "optimal" design, but to understand sensitivity to assumptions and trade-offs among competing choices. Used in this way, simulation complements substantive knowledge and design-based reasoning by making the consequences of analytic decisions explicit before data are collected. The architecture of the **simglm** package is designed to reflect this framework, allowing researchers to modify simulation arguments and analysis models independently to identify where designs are most sensitive to failure.

Bibliography

Austin, P. C. (2011). An introduction to propensity score methods for reducing the effects of confounding in observational studies. *Multivariate Behavioral Research*, 46(3):399–424.

Bates, D., Mächler, M., Bolker, B., and Walker, S. (2015). Fitting linear mixed-effects models using lme4. *Journal of Statistical Software*, 67(1):1–48.

Bengtsson, H. (2021). A unifying framework for parallel and distributed processing in R using futures. *The R Journal*, 13(2):208–227.

Blom, M. M., Zarit, S. H., Groot Zwaaftink, R. B., Cuijpers, P., and Pot, A. M. (2015). Effectiveness of an internet intervention for family caregivers of people with dementia: results of a randomized controlled trial. *PloS One*, 10(2):e0116622.

Bloom, H. S. (1995). Minimum detectable effects: A simple way to report the statistical power of experimental designs. *Evaluation Review*, 19(5):547–556.

Card, D. and Krueger, A. B. (1994). Minimum wages and employment: A case study of the fast-food industry in new jersey and pennsylvania. *The American Economic Review*, 84(4):772–793.

Card, D., Lee, D. S., Pei, Z., and Weber, A. (2015). Inference on causal effects in a generalized regression kink design. *Econometrica*, 83(6):2453–2483.

Cook, T. D., Campbell, D. T., and Shadish, W. (2002). *Experimental and quasi-experimental designs for generalized causal inference*. Houghton Mifflin Boston.

Cunningham, S. (2021). *Causal inference: The mixtape*. Yale university press.

Dempster, A. P., Rubin, D. B., and Tsutakawa, R. K. (1981). Estimation in covariance components models. *Journal of the American Statistical Association*, 76(374):341–353.

Diggle, P. J. (2002). *Analysis of longitudinal data*. Oxford University Press.

Dong, N., & Maynard, R. A. (2013). PowerUp!: A tool for calculating minimum detectable effect sizes and minimum required sample sizes for experimental and quasi-experimental design studies. *Journal of Research on Educational Effectiveness*, 6(1), 24–67. doi: 10.1080/19345747.2012.673143.

Duchini, E. (2017). Is college remedial education a worthy investment? new evidence from a sharp regression discontinuity design. *Economics of Education Review*, 60:36–53.

Enders, C. K. (2022). *Applied missing data analysis.* Guilford Publications.

Etxeberria, I., Salaberria, K., and Gorostiaga, A. (2021). Online support for family caregivers of people with dementia: a systematic review and meta-analysis of rcts and quasi-experimental studies. *Aging & Mental Health*, 25(7):1165–1180.

Faraway, J. J. (2004). *Linear models with R.* Chapman and Hall/CRC.

Faul, F., Erdfelder, E., Lang, A.-G., and Buchner, A. (2007). G* power 3: A flexible statistical power analysis program for the social, behavioral, and biomedical sciences. *Behavior Research Methods*, 39(2):175–191.

Feinstein, A. R. (1998). P-values and confidence intervals: two sides of the same unsatisfactory coin. *Journal of Clinical Epidemiology*, 51(4):355–360.

Fitzmaurice, G. M., Laird, N. M., and Ware, J. H. (2012). *Applied longitudinal analysis*, volume 998. John Wiley & Sons.

Fox, J. (2015). *Applied regression analysis and generalized linear models.* Sage Publications.

Gelman, A. (2006). Multilevel (hierarchical) modeling: what it can and cannot do. *Technometrics*, 48(3):432–435.

Gelman, A. and Carlin, J. (2014). Beyond power calculations: Assessing type s (sign) and type m (magnitude) errors. *Perspectives on Psychological Science*, 9(6):641–651.

Gelman, A., Hill, J., and Vehtari, A. (2020). *Regression and other stories.* Cambridge University Press.

Goldstein, H. (2011). *Multilevel statistical models.* John Wiley & Sons.

Goodman, S. (2008). A dirty dozen: twelve p-value misconceptions. In *Seminars in hematology*, volume 45, pages 135–140. Elsevier.

Goodman-Bacon, A. (2021). Difference-in-differences with variation in treatment timing. *Journal of Econometrics*, 225(2):254–277.

Graham, J. W. (2009). Missing data analysis: Making it work in the real world. *Annual Review of Psychology*, 60(1):549–576.

Greenland, S., Senn, S. J., Rothman, K. J., Carlin, J. B., Poole, C., Goodman, S. N., and Altman, D. G. (2016). Statistical tests, p values, confidence intervals, and power: a guide to misinterpretations. *European Journal of Epidemiology*, 31:337–350.

Gruber, J. (1994). The Incidence of Mandated Maternity Benefits. *The American Economic Review*, 84(3), 622–641. http://www.jstor.org/stable/2118071

Halekoh, U., Højsgaard, S., and Yan, J. (2006). The R package geepack for generalized estimating equations. *Journal of Statistical Software*, 15/2:1–11.

Hayes, R. J. and Moulton, L. H. (2017). *Cluster randomised trials.* Chapman and Hall/CRC.

Hernán, M. A. and Robins, J. M. (2020). *Causal Inference: What If.* Chapman and Hall/CRC.

Hoenig, J. M. and Heisey, D. M. (2001). The abuse of power: the pervasive fallacy of power calculations for data analysis. *The American Statistician*, 55(1):19–24.

Huntington-Klein, N. (2021). *The effect: An introduction to research design and causality.* Chapman and Hall/CRC.

Imbens, G. W. and Rubin, D. B. (2015). *Causal inference in statistics, social, and biomedical sciences.* Cambridge University Press.

Jacob, B. A. and Lefgren, L. (2004). Remedial education and student achievement: A regression-discontinuity analysis. *Review of economics and statistics*, 86(1):226–244.

Kent, D. M., Rothwell, P. M., Ioannidis, J. P., Altman, D. G., and Hayward, R. A. (2010). Assessing and reporting heterogeneity in treatment effects in clinical trials: a proposal. *Trials*, 11(1):85.

Lawson, J. (2014). *Design and Analysis of Experiments with R.* CRC press.

LeBeau, B. (2025). *simglm: Simulate Models Based on the Generalized Linear Model.* R package version 0.9.30.

LeBeau, B., Harwell, M., Monson, D., Dupuis, D., Medhanie, A., and Post, T. R. (2012). Student and high-school characteristics related to completing a science, technology, engineering or mathematics (STEM) major in college. *Research in Science & Technological Education*, 30(1):17–28.

Levine, M. and Ensom, M. H. (2001). Post hoc power analysis: an idea whose time has passed? *Pharmacotherapy: The Journal of Human Pharmacology and Drug Therapy*, 21(4):405–409.

Lewis-Beck, C. and Lewis-Beck, M. (2015). *Applied regression: An introduction*, volume 22. Sage Publications.

Li, F., Morgan, K. L., and Zaslavsky, A. M. (2018). Balancing covariates via propensity score weighting. *Journal of the American Statistical Association*, 113(521):390–400.

Liang, K.-Y. and Zeger, S. L. (1986). Longitudinal data analysis using generalized linear models. *Biometrika*, 73(1):13–22.

Little, R. J. (1988). A test of missing completely at random for multivariate data with missing values. *Journal of the American Statistical Association*, 83(404):1198–1202.

Little, R. J. and Rubin, D. B. (2019). *Statistical analysis with missing data.* John Wiley & Sons.

Little, T. D., Bovaird, J. A., and Card, N. A. (2007). *Modeling contextual effects in longitudinal studies.* Routledge.

Long, J. S. and Ervin, L. H. (2000). Using heteroscedasticity consistent standard errors in the linear regression model. *The American Statistician*, 54(3):217–224.

Miller, S., Johnson, N., and Wherry, L. R. (2021). Medicaid and mortality: New evidence from linked survey and administrative data. *The Quarterly Journal of Economics*, 136(3):1783–1829.

Morgan, S. L. and Winship, C. (2014). *Counterfactuals and causal inference: Methods and principles for social research.* Cambridge University Press.

Morris, T. P., White, I. R., and Crowther, M. J. (2019). Using simulation studies to evaluate statistical methods. *Statistics in Medicine*, 38(11):2074–2102.

Mosteller, F. and Boruch, R. F. (2002). *Evidence matters: Randomized trials in education research.* Brookings Institution Press.

Neuman, S. B., Newman, E. H., and Dwyer, J. (2011). Educational effects of a vocabulary intervention on preschoolers' word knowledge and conceptual development: A cluster-randomized trial. *Reading Research Quarterly*, 46(3):249–272.

O'Keefe, D. J. (2007). Brief report: Post hoc power, observed power, a priori power, retrospective power, prospective power, achieved power: Sorting out appropriate uses of statistical power analyses. *Communication Methods and Measures*, 1(4):291–299.

Pearl, J. (2009). *Causality.* Cambridge university press.

Raudenbush, S., Spybrook, J., Bloom, H., Congdon, R., Hill, C., and Martínez, A. (2011). *Optimal Design Software for Multi-level and Longitudinal Research.* Version 3.01.

Raudenbush, S. W. (1997). Statistical analysis and optimal design for cluster randomised trials. *Psychological Methods*, 2(2):173–185.

Raudenbush, S. W. and Bryk, A. S. (2002). *Hierarchical linear models: Applications and data analysis methods*, volume 1. sage.

Robinson, D., Hayes, A., and Couch, S. (2022). *broom: Convert Statistical Objects into Tidy Tibbles*. R package version 1.0.0.

Rosenbaum, P. R. and Rubin, D. B. (1983). The central role of the propensity score in observational studies for causal effects. *Biometrika*, 70(1):41–55.

Royston, P., Altman, D. G., and Sauerbrei, W. (2006). Dichotomizing continuous predictors in multiple regression: a bad idea. *Statistics in medicine*, 25(1):127–141.

Rubin, D. B. (1974). Estimating causal effects of treatments in randomized and nonrandomized studies. *Journal of Educational Psychology*, 66(5):688.

Rubin, D. B. (1976). Inference and missing data. *Biometrika*, 63(3):581–592.

Schafer, J. L. and Graham, J. W. (2002). Missing data: our view of the state of the art. *Psychological Methods*, 7(2):147.

Senn, S. (2006). Change from baseline and analysis of covariance revisited. *Statistics in Medicine*, 25(24):4334–4344.

Slocum, T. A., Pinkelman, S. E., Joslyn, P. R., and Nichols, B. (2022). Threats to internal validity in multiple-baseline design variations. *Perspectives on Behavior Science*, 45(3):619–638.

Smith, J. D. (2012). Single-case experimental designs: a systematic review of published research and current standards. *Psychological Methods*, 17(4):510.

Stuart, E. A. (2010). Matching methods for causal inference: A review and a look forward. *Statistical Science: A Review Journal of the Institute of Mathematical Statistics*, 25(1):1.

Sullivan, A. L. and Field, S. (2013). Do preschool special education services make a difference in kindergarten reading and mathematics skills?: A propensity score weighting analysis. *Journal of School Psychology*, 51(2):243–260.

Turner, D., Schünemann, H. J., Griffith, L. E., Beaton, D. E., Griffiths, A. M., Critch, J. N., and Guyatt, G. H. (2010). The minimal detectable change cannot reliably replace the minimal important difference. *Journal of Clinical Epidemiology*, 63(1):28–36.

Wakefield, J. (2003). Sensitivity analyses for ecological regression. *Biometrics*, 59(1):9–17.

Wasserstein, R. L., & Lazar, N. A. (2016). The ASA Statement on p-Values: Context, Process, and Purpose. *The American Statistician*, 70(2), 129–133. https://doi.org/10.1080/00031305.2016.1154108

Wickham, H., Averick, M., Bryan, J., Chang, W., McGowan, L. D., François, R., Grolemund, G., Hayes, A., Henry, L., Hester, J., et al. (2019). Welcome to the tidyverse. *Journal of Open Source Software*, 4(43):1686.

Grolemund, G., & Wickham, H. (2017). *R for Data Science.* O'Reilly Media.

Xie, Y. (2015). *Dynamic Documents with R and knitr.* Chapman and Hall/CRC, Boca Raton, Florida, 2nd edition. ISBN 978-1498716963.

Xie, Y. (2022). *bookdown: Authoring Books and Technical Documents with R Markdown.* R package version 0.27.

Zeger, S. L., Liang, K.-Y., and Albert, P. S. (1988). Models for longitudinal data: a generalized estimating equation approach. *Biometrics*, pages 1049–1060.

Index

For Product Safety Concerns and Information please contact our EU representative GPSR@taylorandfrancis.com
Taylor & Francis Verlag GmbH, Kaufingerstraße 24, 80331 München, Germany

www.ingramcontent.com/pod-product-compliance
Ingram Content Group UK Ltd.
Pitfield, Milton Keynes, MK11 3LW, UK
UKHW020158200926
475702UK00004B/522
* 9 7 8 1 0 3 2 0 1 2 5 0 6 *